Yearbook of
Astronomy 2020

Front Cover: This mosaic image of asteroid Bennu is composed of 12 PolyCam (robotic camera system) images collected on 2 December 2018 by the OSIRIS-REx spacecraft from a range of 24 kilometres. The image was obtained at a 50° phase angle between the spacecraft, asteroid and the Sun, and in it Bennu spans approximately 1,500 pixels in the camera's field of view. More information on the OSIRIS-REx mission to Bennu can be found in the article *Solar System Exploration in 2019* elsewhere in this volume (NASA/OSIRIS-REx/Goddard/University of Arizona)

YEARBOOK OF
ASTRONOMY
2020

EDITED BY

Brian Jones

WHITE OWL

AN IMPRINT OF PEN & SWORD BOOKS LTD.
YORKSHIRE - PHILADELPHIA

First published in Great Britain in 2019 by
Pen & Sword WHITE OWL
An imprint of
Pen & Sword Books Ltd
Yorkshire – Philadelphia

Copyright © Brian Jones, 2019

ISBN 978 1 52675 327 4

The right of Brian Jones to be identified as Author of this work has been
asserted by him in accordance with the Copyright, Designs and Patents
Act 1988.

A CIP catalogue record for this book is available from the British Library.

Typeset in Dante By Mac Style

Printed and bound in India by Replika Press Pvt. Ltd.

Pen & Sword Books Ltd incorporates the Imprints of Pen & Sword Books
Archaeology, Atlas, Aviation, Battleground, Discovery, Family History,
History, Maritime, Military, Naval, Politics, Railways, Select, Transport,
True Crime, Fiction, Frontline Books, Leo Cooper, Praetorian Press,
Seaforth Publishing, Wharncliffe and White Owl.

For a complete list of Pen & Sword titles please contact

PEN & SWORD BOOKS LIMITED
47 Church Street, Barnsley, South Yorkshire, S70 2AS, England
E-mail: enquiries@pen-and-sword.co.uk
Website: www.pen-and-sword.co.uk

or

PEN AND SWORD BOOKS
1950 Lawrence Rd, Havertown, PA 19083, USA
E-mail: Uspen-and-sword@casematepublishers.com

Contents

Editor's Foreword

The Yearbook of Astronomy 2020 is the latest edition of what has long been an indispensable publication, the annual appearance of which has been eagerly anticipated by astronomers, both amateur and professional, for well over half a century and which is approaching its Diamond Jubilee edition in 2022. As ever, the Yearbook is aimed at both the armchair astronomer and the active backyard observer. Within its pages you will find a rich blend of information, star charts and guides to the night sky coupled with an interesting mixture of articles which collectively embrace a wide range of topics, ranging from the history of astronomy to the latest results of astronomical research; space exploration to observational astronomy; and our own celestial neighbourhood out to the farthest reaches of space.

The *Monthly Star Charts* have been compiled by David Harper and show the night sky as seen throughout the year. Two sets of twelve charts have been provided, one set for observers in the Northern Hemisphere and one for those in the Southern Hemisphere. Between them, each pair of charts depicts the entire sky as two semi-circular half-sky views, one looking north and the other looking south.

Summaries of the observing conditions for each of the planets, as well as a calendar of significant Solar System events occurring throughout the year, are included in the Yearbook, with *The Planets in 2020* and *Some Events in 2020*. The ongoing process of improving and updating what the Yearbook offers to its readers is continued in the 2020 edition with the introduction of apparition diagrams for Mercury and Venus and finder charts for Uranus and Neptune. These are described in more detail in *Using the Yearbook of Astronomy as an Observing Guide*.

Lists of *Phases of the Moon in 2020* and *Eclipses in 2020* are also provided. Suggestions from readers for further improvements and additions to the Yearbook are always welcomed. After all, the book is written for you …

The *Monthly Sky Notes* have been compiled by Lynne Marie Stockman and give details of the positions and visibility of the planets throughout 2020. Each section of the Monthly Sky Notes is accompanied by a short article, the range of which includes items on a wide variety of astronomy-related topics.

The Monthly Sky Notes and Articles section of the book rounds off with *Comets in 2020*, *Minor Planets in 2020* and *Meteor Showers in 2020*, all three written and provided by Neil Norman and all three titles being fairly self-explanatory, describing as they do the occurrence and visibility of examples of these three classes of object during and throughout the year.

In his article *Astronomy in 2019* Rod Hine covers a range of topics, one of which is the continuing work being carried out with the Atacama Large Millimeter/submillimeter Array (ALMA). The exceptionally fine resolution achieved by ALMA is providing astronomers with new and exciting insights shedding light on such diverse topics as the formation of protoplanetary disks and magnetic fields in jets from newly-formed stars.

This is followed by *Solar System Exploration in 2019* in which Peter Rea provides information on a wide range of space missions spanning our region of space. These include a look forward to the ambitious mission to study the Sun being undertaken by the Parker Solar Probe, launched in August 2018 with the objective of making repeated close passes of the Sun, and the encounter by the New Horizons probe of the Kuiper Belt object Ultima Thule, located way beyond Pluto and out in the cold and distant depths of our solar system.

In 2020 we celebrate the 300th anniversary of the death of the first Astronomer Royal, the Revd. John Flamsteed. In his article *Anniversaries in 2020* Neil Haggath tells us of the Royal Observatory being established by King Charles II at Greenwich in 1675 and of the subsequent appointment of John Flamsteed as its director and Astronomer Royal, a post he held until his death in 1719. Among the other anniversaries covered in Neil's article is the 150th anniversary of the birth of Eugène Michel Antoniadi, widely acclaimed as one of the greatest of all planetary observers. A highly respected observer of Mars, Antoniadi numbers amongst his achievements the establishment of a system of naming for Martian features, many of the names from Antoniadi's 1929 map of the planet still being in use today.

200 Years of the Royal Astronomical Society by Sian Prosser provides us with an interesting and absorbing look at the history of the Royal Astronomical Society, and the continuing and ever expanding role it has played in the promotion, study and understanding of astronomy and associated disciplines. From its original proposal by schoolmaster and astronomer William Pearson in 1812, and eventual founding and inaugural meeting in January 1820, the Royal Astronomical Society has dedicated itself to the pursuit of astronomy and support of the world wide astronomical community. The way we study the Universe has seen many changes, including the birth of astrophysics and the announcement of the theory of relativity, and during the two centuries that the RAS has been with us, the Society has kept pace with developments and continued to play its part.

This is followed by *The Naming of Stars* by David Harper, who draws our attention to the brilliant first-magnitude star Vega which dominates northern hemisphere skies during summer and autumn. In his article David explores the many names that have been given to Vega in different cultures, and uses it as an example to illustrate the history of star-naming by astronomers from ancient times to the modern day. He explains the role played by the International Astronomical Union in approving star names, and warns against commercial companies that falsely claim the authority to name stars after people for a fee.

You might have photographed the objects that you observe, but have you ever thought about sketching them? An alternative to astrophotography as a way of recording what we see through the telescope is provided by Steve Brown in his article *Astronomical Sketching* in which he takes us through all the methods and techniques you will need to sketch and manually record your astronomical observations.

In his article *Dark Matter and Galaxies* Julian Onions explores the mystery of dark matter and the ongoing research being carried out by astronomers to locate and identify it. Although the idea of dark matter is supported by the majority of astronomers, it hides itself extremely well and our efforts to find it are continuing to lead us nowhere.

As Tracie Heywood informs us in *Eclipsing Binaries*, the story of eclipsing binary stars begins during the 1780s when the young York-based astronomer John Goodricke was carrying out investigations into the brightness changes of the star Beta (β) Persei, better known as Algol. This is without doubt the most famous and widely observed example of this class of variable. It can be a very rewarding experience to visually observe an eclipsing binary, watching the star fade in brightness over a number of hours and then brighten again. As we learn in the article, although some eclipse observations are carried out by professional astronomers, the longer-term monitoring of eclipses has long been dependent on visual observations made by amateur astronomers, leaving the door open for backyard astronomers to make their mark.

Australian astronomer Greg Quicke has learned his stars through a life lived outdoors in the wild Kimberley region of Western Australia, and in *A Perspective on the Aboriginal View of the World* he offers us a wonderful and entertaining insight into Aboriginal astronomy. As we see in the article, being born into a culture that has lived under the stars for generations ensures that you can tune into everything that is around you. The stars themselves become like old friends, and the connections between seemingly unrelated things become increasingly apparent. Greg writes about the Aboriginal perspective of the night sky through experiences gained over half a lifetime living in a place, and around the original people of the land, where the natural ways of life are still quite normal and the world is a part of who and what they are.

Although nowadays many people are aware of the term 'black hole', there was a time when the notion of such an object in space was so outrageous that many astronomers were sure that they simply could not exist! But in *The First Known Black Hole* by regular contributor David M. Harland, we see how astronomers investigating Cygnus X-1 discovered that when the core of a massive star collapses it can indeed create a black hole.

The journey through our region of space of an interstellar visitor that was destined to rewrite the science books is described in the article 'Oumuamua – *Interstellar Interloper* by Neil Norman. Subsequent observation, following the discovery on the night of 19 October 2017 by astronomer Robert J. Weryk of a rapidly moving object in the constellation of Cetus, soon made it apparent that 'Oumuamua was the first unbound object detected travelling through the Solar System. In his article, Neil outlines the research that was carried out into the physical and orbital characteristics of 'Oumuamua, as well as attempting to answer the question as to where the interstellar interloper originally came from.

Bill Leatherbarrow rounds off our articles with his *Forgotten on the Moon* which provides an interesting look at how the history of lunar nomenclature is littered with the names of those who fell by the wayside and whose contributions, large and small, have been denied recognition on the Moon. A notable example quoted is the crater Shapley on the southern boundary of Mare Crisium which commemorates the American astronomer Harlow Shapley, but fails to recognise the contribution made by his wife Martha Betz Shapley who provided invaluable assistance to Shapley's researches. Similarly, the achievements of German astronomer Gottfried Kirch are marked by a small crater north of Archimedes on Mare Imbrium, although the work carried out by his second wife Maria Margaretha Kirch (née Winckelmann), who assisted him in many of his observations, is not represented. Many more examples are quoted and the article certainly gives food for thought ...

The final section of the book starts off with *Some Interesting Variable Stars* by Roger Pickard which contains useful information on variables as well as predictions for timings of minimum brightness of the famous eclipsing binary Algol for 2020. *Some Interesting Double Stars* and *Some Interesting Nebulae, Star Clusters and Galaxies* present a selection of objects for you to seek out in the night sky. The lists included here are by no means definitive and may well omit your favourite celestial targets. If this is the case, please let us know and we will endeavour to include these in future editions of the Yearbook.

The book rounds off with a selection of *Astronomical Organizations*, which lists organizations and associations across the world through which you can further pursue your interest and participation in astronomy (if there are any that we have omitted please let us know) and *Our Contributors*, which contains brief background details of the numerous writers who have contributed to the Yearbook of Astronomy 2020. Finally, the *Glossary* has been expanded a little for 2020 with the addition of new entries of brief but informative explanations for words and terminology used in the Yearbook.

Over time new topics and themes will be introduced into the Yearbook to allow it to keep pace with the increasing range of skills, techniques and observing methods now open to amateur astronomers, this in addition to articles relating to our rapidly-expanding knowledge of the Universe in which we live. There will be an interesting mix, some articles written at a level which will appeal to the casual reader and some of what may be loosely described as at a more academic level. The intention is to fully maintain and continually increase the usefulness and relevance of the Yearbook of Astronomy to the interests of the readership who are, without doubt, the most important aspect of the Yearbook and the reason it exists in the first place.

As ever, grateful thanks are extended to those individuals who have contributed a great deal of time and effort to the Yearbook of Astronomy 2020, not least of which is David Harper, who has provided updated versions of his excellent Monthly Star Charts. These were generated specifically for what has been described as the new generation of the Yearbook of Astronomy, and the charts add greatly to the overall value of the book to star gazers. Equally important are the efforts of Lynne Marie Stockman who has put together the Monthly Sky Notes. Their combined efforts have produced what can justifiably be described as the backbone of the Yearbook of Astronomy.

Also worthy of mention is Garry (Garfield) Blackmore who, as well as preparing many of the illustrations for publication, has produced the artwork for the smaller star charts included with a number of the articles. Thanks also go to Mat Blurton, who has done an excellent job typesetting the Yearbook, and to Jonathan Wright, Emily Robinson, Lori Jones, Janet Brookes and Paul Wilkinson of Pen & Sword Books Ltd without whose combined help, belief and confidence in the Yearbook of Astronomy, this much-loved and iconic publication may well have disappeared for ever.

Brian Jones – Editor
Bradford, West Riding of Yorkshire

February 2019

As many of you will be aware, the future of the Yearbook of Astronomy was under threat following the decision to make the 2016 edition the last. However, the series was rescued, both through the publication of a special 2017 edition (which successfully maintained the continuity of the Yearbook) and a successful search for a publisher to take this iconic publication on and to carry it to even greater heights as the Yearbook approaches its Diamond Jubilee in 2022.

The Yearbook of Astronomy 2017 was a limited edition, although copies are still available to purchase. It should be borne in mind that you would not be obtaining the 2017 edition as a current guide to the night sky, but as the landmark edition of the Yearbook of Astronomy which fulfilled its purpose of keeping the series alive, and which heralded in the new generation of this highly valued and treasured publication. You can order your copy of the 2017 edition at **www.starlight-nights.co.uk/subscriber-2017-yearbook-astronomy**

Preface

The information given in this edition of the Yearbook of Astronomy is in narrative form. The positions of the planets given in the Monthly Sky Notes often refer to the constellations in which they lie at the time. These can be found on the star charts which collectively show the whole sky via two charts depicting the northern and southern circumpolar stars and forty-eight charts depicting the main stars and constellations for each month of the year. The northern and southern circumpolar charts show the stars that are within 45° of the two celestial poles, while the monthly charts depict the stars and constellations that are visible throughout the year from Europe and North America or from Australia and New Zealand. The monthly charts overlap the circumpolar charts. Wherever you are on the Earth, you will be able to locate and identify the stars depicted on the appropriate areas of the chart(s).

There are numerous star atlases available that offer more detailed information, such as *Sky & Telescope's POCKET SKY ATLAS* and *Norton's STAR ATLAS and Reference Handbook* to name but a couple. In addition, more precise information relating to planetary positions and so on can be found in a number of publications, a good example of which is *The Handbook of the British Astronomical Association*, as well as many of the popular astronomy magazines such as the British monthly periodicals *Sky at Night* and *Astronomy Now* and the American monthly magazines *Astronomy* and *Sky & Telescope*.

About Time

Before the late 18th century, the biggest problem affecting mariners sailing the seas was finding their position. Latitude was easily determined by observing the altitude of the pole star above the northern horizon. Longitude, however, was far more difficult to measure. The inability of mariners to determine their longitude often led to them getting lost, and on many occasions shipwrecked. To address this problem King Charles II established the Royal Observatory at Greenwich in 1675 and from here, Astronomers Royal began the process of measuring and cataloguing the stars as they passed due south across the Greenwich meridian.

Now mariners only needed an accurate timepiece (the chronometer invented by Yorkshire-born clockmaker John Harrison) to display GMT (Greenwich Mean Time). Working out the local standard time onboard ship and subtracting this from GMT gave the ship's longitude (west or east) from the Greenwich meridian. Therefore mariners always knew where they were at sea and the longitude problem was solved.

Astronomers use a time scale called Universal Time (UT). This is equivalent to Greenwich Mean Time and is defined by the rotation of the Earth. The Yearbook of Astronomy gives all times in UT rather than in the local time for a particular city or country. Times are expressed using the 24-hour clock, with the day beginning at midnight, denoted by 00:00. Universal Time (UT) is related to local mean time by the formula:

Local Mean Time = UT – west longitude

In practice, small differences in longitude are ignored and the observer will use local clock time which will be the appropriate Standard (or Zone) Time. As the formula indicates, places in west longitude will have a Standard Time slow on UT, while those in east longitude will have a Standard Time fast on UT. As examples we have:

Standard Time in

New Zealand	UT +12 hours
Victoria, NSW	UT +10 hours
Western Australia	UT + 8 hours
South Africa	UT + 2 hours
British Isles	UT
Eastern Standard Time	UT −5 hours
Central Standard Time	UT −6 hours
Pacific Standard Time	UT −8 hours

During the periods when Summer Time (also called Daylight Saving Time) is in use, one hour must be added to Standard Time to obtain the appropriate Summer/Daylight Saving Time. For example, Pacific Daylight Time is UT −7 hours.

Using the Yearbook of Astronomy as an Observing Guide

Notes on the Monthly Star Charts

The star charts on the following pages show the night sky throughout the year. There are two sets of charts, one for use by observers in the Northern Hemisphere and one for those in the Southern Hemisphere. The first set is drawn for latitude 52°N and can be used by observers in Europe, Canada and most of the United States. The second set is drawn for latitude 35°S and show the stars as seen from Australia and New Zealand. Twelve pairs of charts are provided for each of these latitudes.

Each pair of charts shows the entire sky as two semi-circular half-sky views, one looking north and the other looking south. A given pair of charts can be used at different times of year. For example, chart 1 shows the night sky at midnight on 21 December, but also at 2am on 21 January, 4am on 21 February and so forth. The accompanying table will enable you to select the correct chart for a given month and time of night. The caption next to each chart also lists the dates and times of night for which it is valid.

The charts are intended to help you find the more prominent constellations and other objects of interest mentioned in the monthly observing notes. To avoid the charts becoming too crowded, only stars of magnitude 4.5 or brighter are shown. This corresponds to stars that are bright enough to be seen from any dark suburban garden on a night when the Moon is not too close to full phase.

Each constellation is depicted by joining selected stars with lines to form a pattern. There is no official standard for these patterns, so you may occasionally find different patterns used in other popular astronomy books for some of the constellations.

Any map projection from a sphere onto a flat page will by necessity contain some distortions. This is true of star charts as well as maps of the Earth. The distortion on the half-sky charts is greatest near the semi-circular boundary of each chart, where it may appear to stretch constellation patterns out of shape.

The charts also show selected deep-sky objects such as galaxies, nebulae and star clusters. Many of these objects are too faint to be seen with the naked eye, and you will need binoculars or a telescope to observe them. Please refer to the table of deep-sky objects for more information.

Planetary Apparition Diagrams

The diagrams of the apparitions of Mercury and Venus show the position of the respective planet in the sky at the moment of sunrise or sunset throughout the entire apparition. Two sets of positions are plotted on each chart: for latitude 52° North (blue line) and for latitude 35° South (red line). A thin dotted line denotes the portion of the apparition which falls outside the year covered by this edition of the Yearbook. A white dot indicates the position of Venus on the first day of each month, or of Mercury on the first, eleventh and 21st of the month. The day of greatest elongation (GE) is also marked by a white dot. Note that the dots do NOT indicate the magnitude of the planet.

The finder charts for Uranus and Neptune show the paths of the planets throughout the year. The position of each planet is indicated at opposition and at stationary points, as well as the start and end of the year and on the 1ˢᵗ of April, July and October where these dates do not fall too close to an event that is already marked. On the Uranus chart, stars are shown to magnitude 8; on the Neptune chart, the limiting magnitude is 10. In both cases, this is approximately two magnitudes fainter than the planet itself. Right Ascension and Declination scales are shown for the epoch J2000 to allow comparison with modern star charts.

Selecting the Correct Charts

The table below shows which of the charts to use for particular dates and times throughout the year and will help you to select the correct pair of half-sky charts for any combination of month and time of night.

The Earth takes 23 hours 56 minutes (and 4 seconds) to rotate once around its axis with respect to the fixed stars. Because this is around four minutes shorter than a full 24 hours, the stars appear to rise and set about 4 minutes earlier on each successive day, or around an hour earlier each fortnight. Therefore, as well as showing the stars at 10pm (22h in 24-hour notation) on 21 January, chart 1 also depicts the sky at 9pm (21h) on 6 February, 8pm (20h) on 21 February and 7pm (19h) on 6 March.

The times listed do not include summer time (daylight saving time), so if summer time is in force you must subtract one hour to obtain standard time (GMT if you are in the United Kingdom) before referring to the chart. For example, to find the correct chart for mid-September in the northern hemisphere at 3am summer time, first of all subtract one hour to obtain 2am (2h) standard time. Then you can consult the table, where you will find that you should use chart 11.

The table does not indicate sunrise, sunset or twilight. In northern temperate latitudes, the sky is still light at 18h and 6h from April to September, and still light at 20h and 4h from May to August. In Australia and New Zealand, the sky is still light at 18h and 6h from October to March, and in twilight (with only bright stars visible) at 20h and 04h from November to January.

Local Time	18h	20h	22h	0h	2h	4h	6h
January	11	12	1	2	3	4	5
February	12	1	2	3	4	5	6
March	1	2	3	4	5	6	7
April	2	3	4	5	6	7	8
May	3	4	5	6	7	8	9
June	4	5	6	7	8	9	10
July	5	6	7	8	9	10	11
August	6	7	8	9	10	11	12
September	7	8	9	10	11	12	1
October	8	9	10	11	12	1	2
November	9	10	11	12	1	2	3
December	10	11	12	1	2	3	4

Legend to the Star Charts

STARS		DEEP-SKY OBJECTS	
Symbol	Magnitude	Symbol	Type of object
•	0 or brighter	✳	Open star cluster
•	1	○	Globular star cluster
•	2	□	Nebula
•	3	▣	Cluster with nebula
•	4	○	Planetary nebula
·	5	◠	Galaxy
✦	Double star		Magellanic Clouds
◉	Variable star		

Star Names

There are over 200 stars with proper names, most of which are of Roman, Greek or Arabic origin although only a couple of dozen or so of these names are used regularly. Examples include Arcturus in Boötes, Castor and Pollux in Gemini and Rigel in Orion.

A system whereby Greek letters were assigned to stars was introduced by the German astronomer and celestial cartographer Johann Bayer in his star atlas Uranometria, published in 1603. Bayer's system is applied to the brighter stars within any particular constellation, which are given a letter from the Greek alphabet followed by the genitive case of the constellation in which the star is located. This genitive case is simply the Latin form meaning 'of' the constellation. Examples are the stars Alpha Boötis and Beta Centauri which translate literally as 'Alpha of Boötes' and 'Beta of the Centaur'.

As a general rule, the brightest star in a constellation is labelled Alpha (α), the second brightest Beta (β), and the third brightest Gamma (γ) and so on, although there are some constellations where the system falls down. An example is Gemini where the principal star (Pollux) is designated Beta Geminorum, the second brightest (Castor) being known as Alpha Geminorum.

There are only 24 letters in the Greek alphabet, the consequence of which was that the fainter naked eye stars needed an alternative system of classification. The system in popular use is that devised by the first Astronomer Royal John

Flamsteed in which the stars in each constellation are listed numerically in order from west to east. Although many of the brighter stars within any particular constellation will have both Greek letters and Flamsteed numbers, the latter are generally used only when a star does not have a Greek letter.

The Greek Alphabet

α	Alpha	ι	Iota	ρ	Rho
β	Beta	κ	Kappa	σ	Sigma
γ	Gamma	λ	Lambda	τ	Tau
δ	Delta	μ	Mu	υ	Upsilon
ε	Epsilon	ν	Nu	φ	Phi
ζ	Zeta	ξ	Xi	χ	Chi
η	Eta	o	Omicron	ψ	Psi
θ	Theta	π	Pi	ω	Omega

The Names of the Constellations

On clear, dark, moonless nights, the sky seems to teem with stars although in reality you can never see more than a couple of thousand or so at any one time when looking with the unaided eye. Each and every one of these stars belongs to a particular constellation, although the constellations that we see in the sky, and which grace the pages of star atlases, are nothing more than chance alignments. The stars that make up the constellations are often situated at vastly differing distances from us and only appear close to each other, and form the patterns that we see, because they lie in more or less the same direction as each other as seen from Earth.

A large number of the constellations are named after mythological characters, and were given their names thousands of years ago. However, those star groups lying close to the south celestial pole were discovered by Europeans only during the last few centuries, many of these by explorers and astronomers who mapped the stars during their journeys to lands under southern skies. This resulted in many of the newer constellations having modern-sounding names, such as Octans (the Octant) and Microscopium (the Microscope), both of which were devised by the French astronomer Nicolas Louis De La Caille during the early 1750s.

Over the centuries, many different suggestions for new constellations have been put forward by astronomers who, for one reason or another, felt the need to add new groupings to star charts and to fill gaps between the traditional constellations. Astronomers drew up their own charts of the sky, incorporating their new groups into them. A number of these new constellations had cumbersome names, notable examples including Officina Typographica (the Printing Shop) introduced by the German astronomer Johann Bode in 1801; Sceptrum Brandenburgicum (the Sceptre of Brandenburg) introduced by the German astronomer Gottfried Kirch in 1688; Taurus Poniatovii (Poniatowski's Bull) introduced by the Polish-Lithuanian astronomer Martin Odlanicky Poczobut in 1777; and Quadrans Muralis (the Mural Quadrant) devised by the French astronomer Joseph-Jerôme de Lalande in1795. Although these have long since been rejected, the latter has been immortalised by the annual Quadrantid meteor shower, the radiant of which lies in an area of sky formerly occupied by Quadrans Muralis.

During the 1920s the International Astronomical Union (IAU) systemised matters by adopting an official list of 88 accepted constellations, each with official spellings and abbreviations. Precise boundaries for each constellation were then drawn up so that every point in the sky belonged to a particular constellation.

The abbreviations devised by the IAU each have three letters which in the majority of cases are the first three letters of the constellation name, such as AND for Andromeda, EQU for Equuleus, HER for Hercules, ORI for Orion and so on. This trend is not strictly adhered to in cases where confusion may arise. This happens with the two constellations Leo (abbreviated LEO) and Leo Minor (abbreviated LMI). Similarly, because Triangulum (TRI) may be mistaken for Triangulum Australe, the latter is abbreviated TRA. Other instances occur with Sagitta (SGE) and Sagittarius (SGR) and with Canis Major (CMA) and Canis Minor (CMI) where the first two letters from the second names of the constellations are used. This is also the case with Corona Australis (CRA) and Corona Borealis (CRB) where the first letter of the second name of each constellation is incorporated. Finally, mention must be made of Crater (CRT) which has been abbreviated in such a way as to avoid confusion with the aforementioned CRA (Corona Australis).

The table shown on the following pages contains the name of each of the 88 constellations together with the translation and abbreviation of the constellation name. The constellations depicted on the monthly star charts are identified with their abbreviations rather than the full constellation names.

The Constellations

Andromeda	Andromeda	AND	Delphinus	The Dolphin	DEL
Antlia	The Air Pump	ANT	Dorado	The Goldfish	DOR
Apus	The Bird of Paradise	APS	Draco	The Dragon	DRA
			Equuleus	The Foal	EQU
Aquarius	The Water Carrier	AQR	Eridanus	The River	ERI
Aquila	The Eagle	AQL	Fornax	The Furnace	FOR
Ara	The Altar	ARA	Gemini	The Twins	GEM
Aries	The Ram	ARI	Grus	The Crane	GRU
Auriga	The Charioteer	AUR	Hercules	Hercules	HER
Boötes	The Herdsman	BOO	Horologium	The Pendulum Clock	HOR
Caelum	The Graving Tool	CAE			
Camelopardalis	The Giraffe	CAM	Hydra	The Water Snake	HYA
Cancer	The Crab	CNC	Hydrus	The Lesser Water Snake	HYI
Canes Venatici	The Hunting Dogs	CVN	Indus	The Indian	IND
Canis Major	The Great Dog	CMA	Lacerta	The Lizard	LAC
Canis Minor	The Little Dog	CMI	Leo	The Lion	LEO
Capricornus	The Goat	CAP	Leo Minor	The Lesser Lion	LMI
Carina	The Keel	CAR	Lepus	The Hare	LEP
Cassiopeia	Cassiopeia	CAS	Libra	The Scales	LIB
Centaurus	The Centaur	CEN	Lupus	The Wolf	LUP
Cepheus	Cepheus	CEP	Lynx	The Lynx	LYN
Cetus	The Whale	CET	Lyra	The Lyre	LYR
Chamaeleon	The Chameleon	CHA	Mensa	The Table Mountain	MEN
Circinus	The Pair of Compasses	CIR	Microscopium	The Microscope	MIC
			Monoceros	The Unicorn	MON
Columba	The Dove	COL	Musca	The Fly	MUS
Coma Berenices	Berenice's Hair	COM	Norma	The Level	NOR
Corona Australis	The Southern Crown	CRA	Octans	The Octant	OCT
Corona Borealis	The Northern Crown	CRB	Ophiuchus	The Serpent Bearer	OPH
Corvus	The Crow	CRV	Orion	Orion	ORI
Crater	The Cup	CRT	Pavo	The Peacock	PAV
Crux	The Cross	CRU	Pegasus	Pegasus	PEG
Cygnus	The Swan	CYG	Perseus	Perseus	PER

Phoenix	The Phoenix	PHE		Sextans	The Sextant	SEX
Pictor	The Painter's Easel	PIC		Taurus	The Bull	TAU
Pisces	The Fish	PSC		Telescopium	The Telescope	TEL
Piscis Austrinus	The Southern Fish	PSA		Triangulum	The Triangle	TRI
Puppis	The Stern	PUP		Triangulum Australe	The Southern Triangle	TRA
Pyxis	The Mariner's Compass	PYX		Tucana	The Toucan	TUC
Reticulum	The Net	RET		Ursa Major	The Great Bear	UMA
Sagitta	The Arrow	SGE		Ursa Minor	The Little Bear	UMI
Sagittarius	The Archer	SGR		Vela	The Sail	VEL
Scorpius	The Scorpion	SCO		Virgo	The Virgin	VIR
Sculptor	The Sculptor	SCL		Volans	The Flying Fish	VOL
Scutum	The Shield	SCT		Vulpecula	The Fox	VUL
Serpens Caput and Cauda	The Serpent	SER				

The Monthly Star Charts

Northern Hemisphere Star Charts

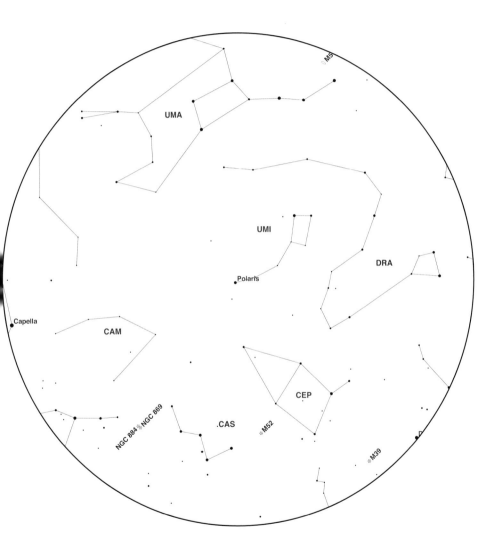

This chart shows stars lying at declinations between +45 and +90 degrees. These constellations are circumpolar for observers in Europe and North America.

1N

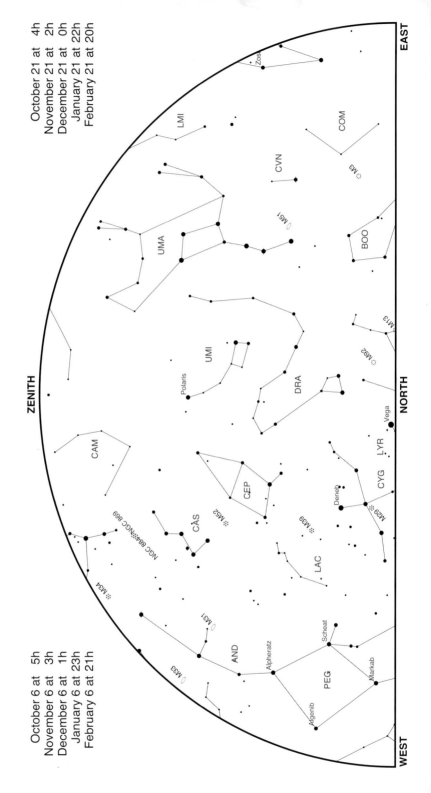

October 21 at 4h
November 21 at 2h
December 21 at 0h
January 21 at 22h
February 21 at 20h

October 6 at 5h
November 6 at 3h
December 6 at 1h
January 6 at 23h
February 6 at 21h

EAST

WEST

ZENITH

NORTH

LMI

COM

CVN

M3

M51

UMA

BOO

M13

M92

UMI

DRA

Polaris

CAM

CEP

CAS

M52

M29

CYG

LYR

Vega

Deneb

M39

LAC

NGC 884/NGC 869

M34

M31

AND

Alpheratz

M33

PEG

Scheat

Markab

Algenib

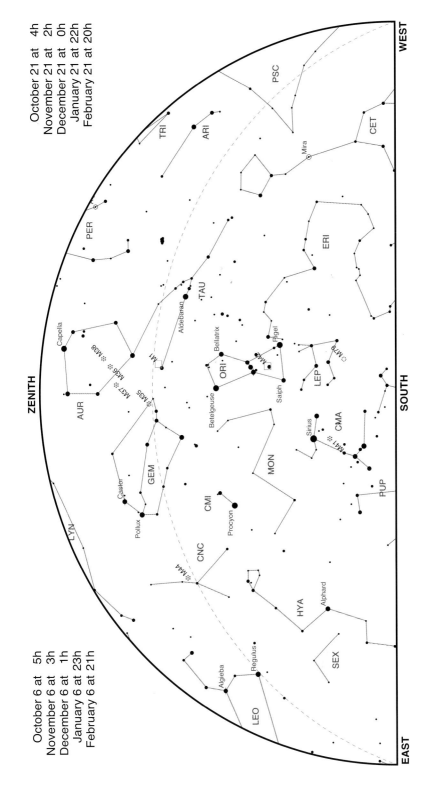

1S

October 21 at 4h
November 21 at 2h
December 21 at 0h
January 21 at 22h
February 21 at 20h

October 6 at 5h
November 6 at 3h
December 6 at 1h
January 6 at 23h
February 6 at 21h

WEST

ZENITH

EAST

SOUTH

PSC

CET

Mira

ARI

TRI

PER

ERI

TAU

Aldebaran

Capella

M38

M36

M37

M35

M1

AUR

Bellatrix

Rigel

ORI

M42

Betelgeuse

Saiph

LEP

M79

Castor

Pollux

GEM

LYN

CMI

Procyon

MON

Sirius

CMA

M41

PUP

CNC

M44

HYA

Alphard

SEX

LEO

Regulus

Algieba

2N

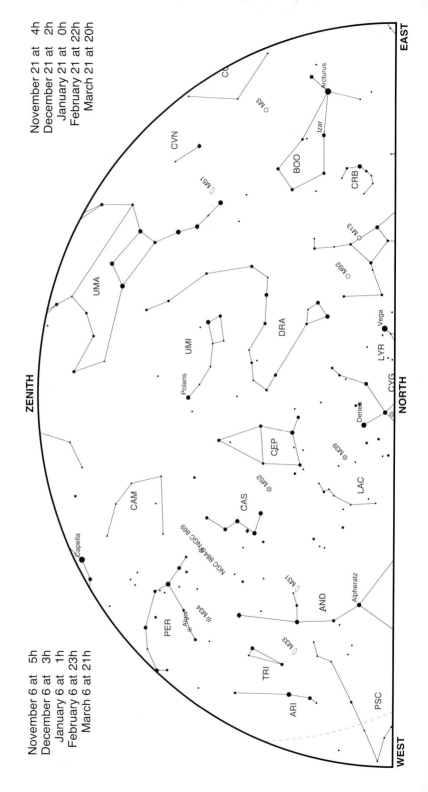

November 21 at 4h
December 21 at 2h
January 21 at 0h
February 21 at 22h
March 21 at 20h

November 6 at 5h
December 6 at 3h
January 6 at 1h
February 6 at 23h
March 6 at 21h

EAST

ZENITH

WEST

NORTH

CVN

CO

M3

Arcturus

Izar

BOO

CRB

M13

M92

UMA

M51

DRA

Vega

LYR

UMI

CYG

Polaris

Deneb

CEP

M39

CAM

M52

CAS

LAC

Capella

NGC 884 NGC 869

Algol

M34

PER

M31

AND

Alpheratz

M33

TRI

ARI

PSC

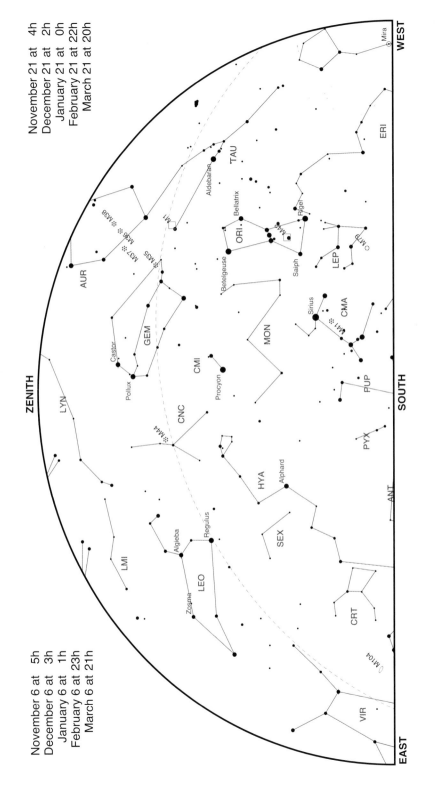

2S

November 21 at 4h
December 21 at 2h
January 21 at 0h
February 21 at 22h
March 21 at 20h

November 6 at 5h
December 6 at 3h
January 6 at 1h
February 6 at 23h
March 6 at 21h

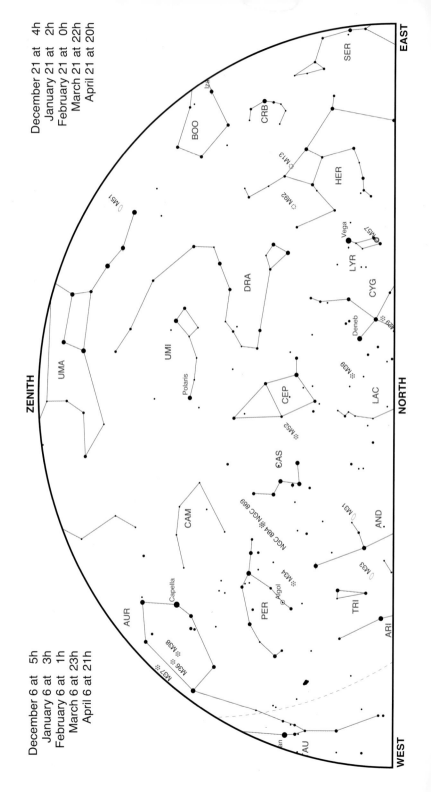

3N

December 6 at 5h
January 6 at 3h
February 6 at 1h
March 6 at 23h
April 6 at 21h

EAST

WEST

NORTH

ZENITH

SER

CRB

BOO

HER

M51

Iz

M13

M92

Vega

M57

LYR

DRA

CYG

Deneb

Deb

UMA

UMI

M39

Polaris

CEP

LAC

M52

CAS

CAM

NGC 884 NGC 869

M34

AND

M31

Capella

AUR

Algol

M33

PER

TRI

M36

M38

ARI

M37

AU

an

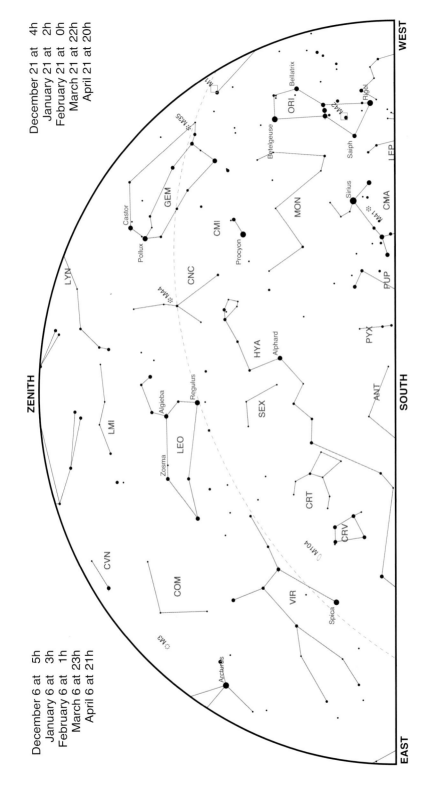

3S

WEST

December 21 at 4h
January 21 at 2h
February 21 at 0h
March 21 at 22h
April 21 at 20h

ZENITH

December 6 at 5h
January 6 at 3h
February 6 at 1h
March 6 at 23h
April 6 at 21h

EAST

SOUTH

4N

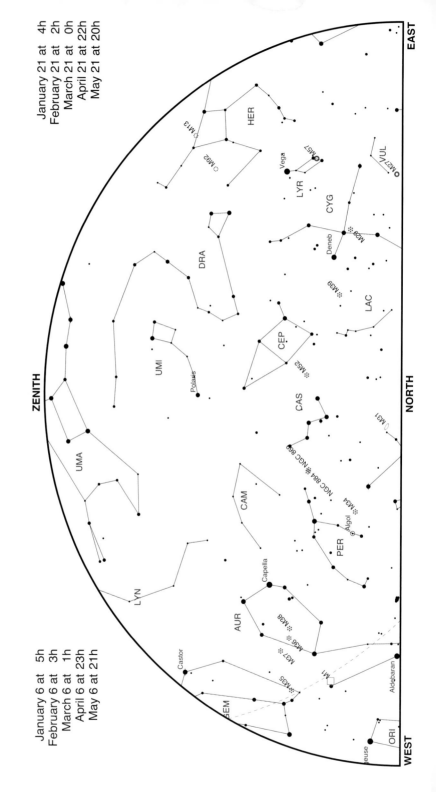

January 21 at 4h
February 21 at 2h
March 21 at 0h
April 21 at 22h
May 21 at 20h

January 6 at 5h
February 6 at 3h
March 6 at 1h
April 6 at 23h
May 6 at 21h

EAST

WEST

NORTH

ZENITH

HER
M13
M92
Vega
M57
LYR
CYG
Deneb
M39
M39
LAC
VUL
M27

DRA
UMI
Polaris
CEP
M52
CAS
NGC 884 NGC 869
M34
Algol
PER
M31

UMA
CAM
LYN
AUR
Capella
M38
M36
M37
M35
GEM
Castor
M1
ORI
Aldebaran
...euse

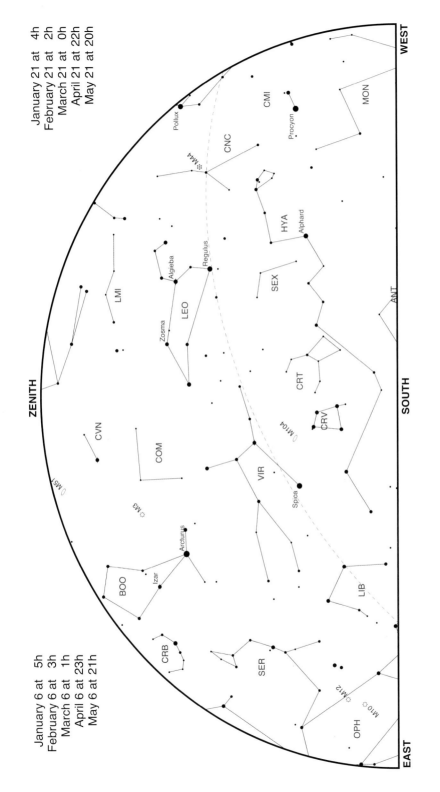

4S

WEST

January 21 at 4h
February 21 at 2h
March 21 at 0h
April 21 at 22h
May 21 at 20h

January 6 at 5h
February 6 at 3h
March 6 at 1h
April 6 at 23h
May 6 at 21h

ZENITH

EAST

SOUTH

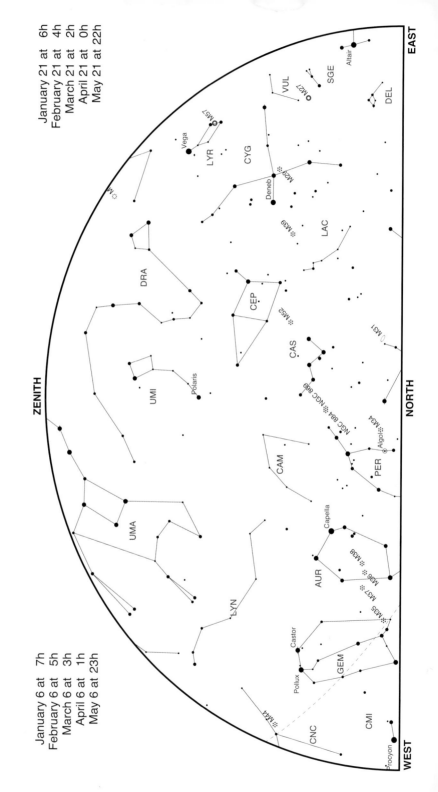

5N

January 21 at 6h
February 21 at 4h
March 21 at 2h
April 21 at 0h
May 21 at 22h

January 6 at 7h
February 6 at 5h
March 6 at 3h
April 6 at 1h
May 6 at 23h

EAST

NORTH

WEST

ZENITH

Altair
DEL
SGE
VUL
M27
M57
Vega
LYR
CYG
Deneb
M29
M39
LAC
CEP
M52
CAS
M31
DRA
NGC 884 NGC 869
M34
Algol
PER
UMI
Polaris
CAM
UMA
Capella
AUR
M38
M36
M37
M35
LYN
Castor
Pollux
GEM
M44
CNC
CMI
Procyon

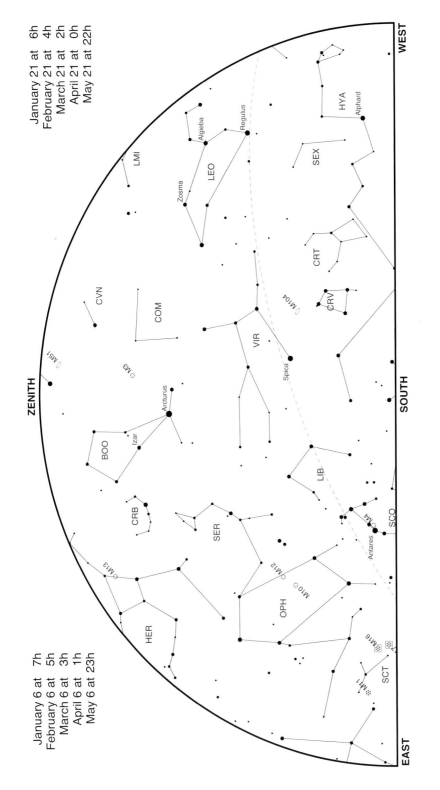

5S

WEST

January 21 at 6h
February 21 at 4h
March 21 at 2h
April 21 at 0h
May 21 at 22h

January 6 at 7h
February 6 at 5h
March 6 at 3h
April 6 at 1h
May 6 at 23h

ZENITH

SOUTH

EAST

HYA
Alphard
SEX
LEO
Algieba
Regulus
Zosma
LMI
CRT
CRV
M104
VIR
Spica
CVN
COM
M3
M51
BOO
Arcturus
Izar
CRB
SER
LIB
HER
M13
OPH
M12
M10
M14
Antares
SCO
M16
M11
SCT

6N

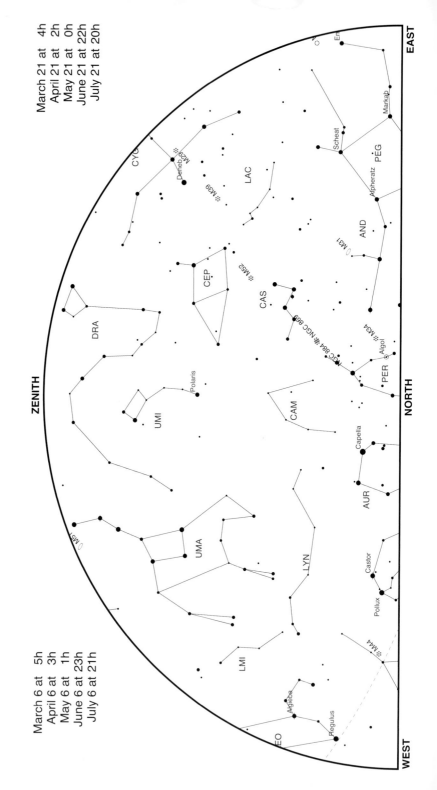

March 21 at 4h
April 21 at 2h
May 21 at 0h
June 21 at 22h
July 21 at 20h

March 6 at 5h
April 6 at 3h
May 6 at 1h
June 6 at 23h
July 6 at 21h

EAST

WEST

NORTH

ZENITH

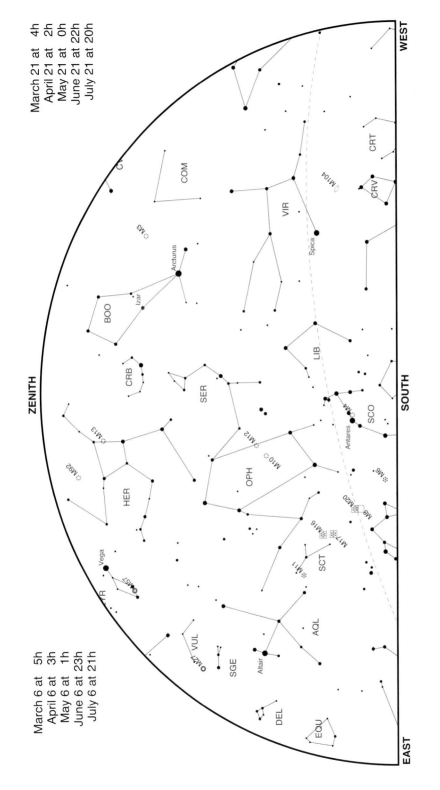

WEST

March 21 at 4h
April 21 at 2h
May 21 at 0h
June 21 at 22h
July 21 at 20h

ZENITH

EAST

March 6 at 5h
April 6 at 3h
May 6 at 1h
June 6 at 23h
July 6 at 21h

SOUTH

COM

BOO
Arcturus
Izar
M3

CRB

SER

VIR
Spica
M104

CRT

CRV

LIB

SCO
Antares
M4
M6

OPH
M10
M12

HER
M13
M92

LYR
Vega
M57
M56

VUL

SGE

AQL
Altair

SCT
M11
M16
M17
M8
M20

DEL

EQU

7N

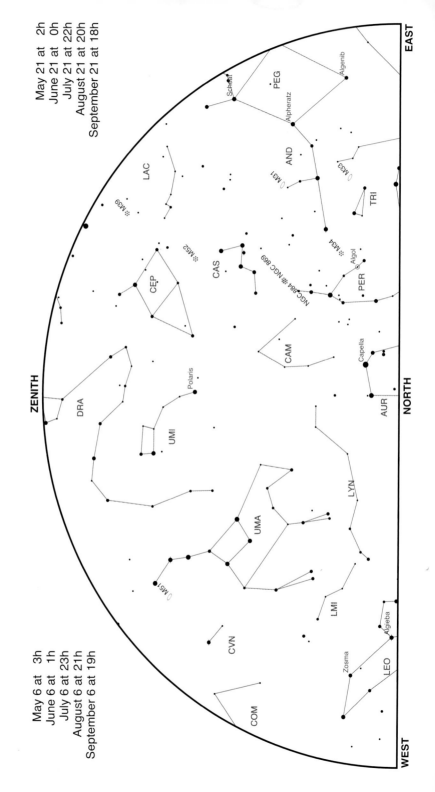

May 21 at 2h
June 21 at 0h
July 21 at 22h
August 21 at 20h
September 21 at 18h

May 6 at 3h
June 6 at 1h
July 6 at 23h
August 6 at 21h
September 6 at 19h

ZENITH

EAST

NORTH

WEST

PEG
Scheat
Algenib
Alpheratz
AND
M31
M33
TRI
M34
Algol
PER
NGC 884 & NGC 869
CAS
M52
CEP
M39
LAC
DRA
Polaris
UMI
CAM
Capella
AUR
LYN
UMA
M51
CVN
COM
LMI
Algieba
Zosma
LEO

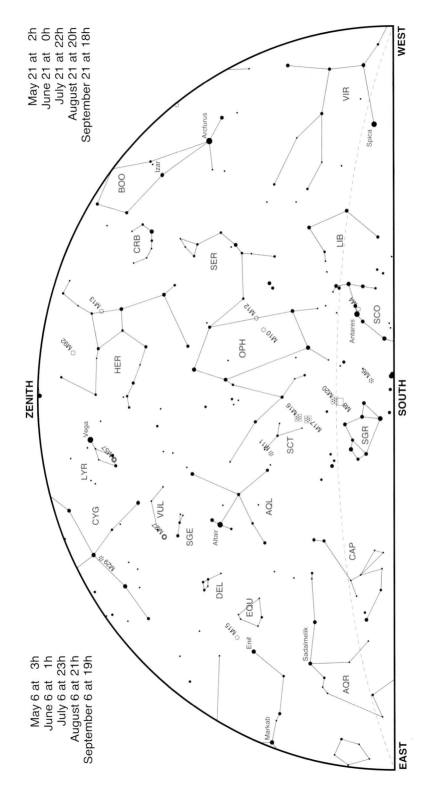

7S

WEST

EAST

SOUTH

ZENITH

May 21 at 2h
June 21 at 0h
July 21 at 22h
August 21 at 20h
September 21 at 18h

May 6 at 3h
June 6 at 1h
July 6 at 23h
August 6 at 21h
September 6 at 19h

VIR
Spica
Arcturus
Izar
BOO
CRB
SER
M13
M92
HER
M12
M10
OPH
LIB
M4
SCO
Antares
M6
Vega
M57
LYR
CYG
M29
VUL
M27
SGE
Altair
AQL
SCT
M11
M17
M16
M20
M8
SGR
M15
DEL
EQU
Enif
CAP
Sadalmelik
Markab
AQR

8N

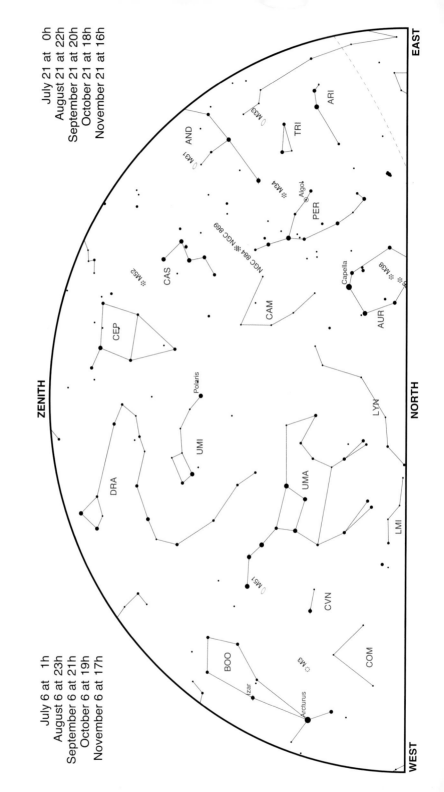

July 21 at 0h
August 21 at 22h
September 21 at 20h
October 21 at 18h
November 21 at 16h

July 6 at 1h
August 6 at 23h
September 6 at 21h
October 6 at 19h
November 6 at 17h

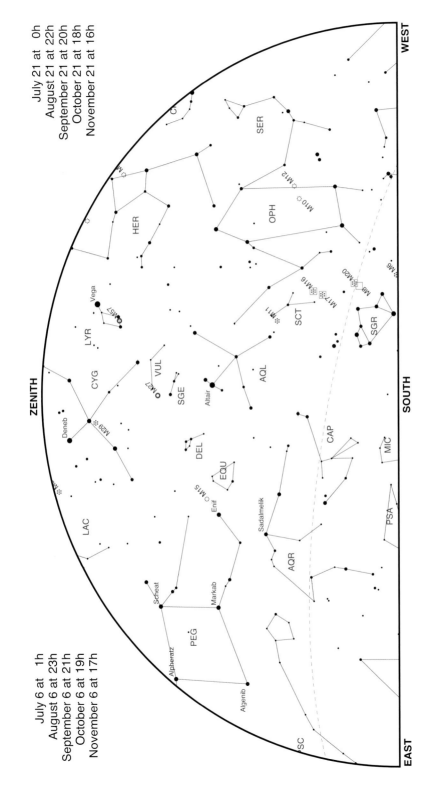

8S

WEST

July 21 at 0h
August 21 at 22h
September 21 at 20h
October 21 at 18h
November 21 at 16h

July 6 at 1h
August 6 at 23h
September 6 at 21h
October 6 at 19h
November 6 at 17h

ZENITH

EAST

SOUTH

SER
HER
OPH
M12
M10
SCT
M16
M17
M11
M6
M20
M8
SGR
LYR
Vega
M57
CYG
Deneb
M29
VUL
M27
SGE
AQL
Altair
DEL
EQU
M15
Enif
LAC
M39
PEG
Scheat
Markab
Alpheratz
Algenib
AQR
Sadalmelik
CAP
MIC
PSA
SC

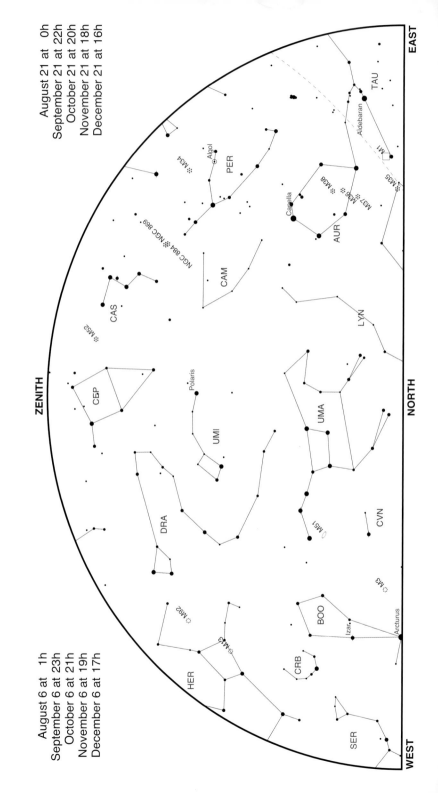

9N

August 6 at 1h
September 6 at 23h
October 6 at 21h
November 6 at 19h
December 6 at 17h

ZENITH

EAST

NORTH

WEST

TAU
Aldebaran
M1
M35
AUR
Capella
M38
M36
M37
PER
Algol
M34
CAM
NGC 884 & NGC 869
CAS
M52
CEP
Polaris
UMI
DRA
LYN
UMA
M51
CVN
M3
BOO
Izar
Arcturus
CRB
M92
M13
HER
SER

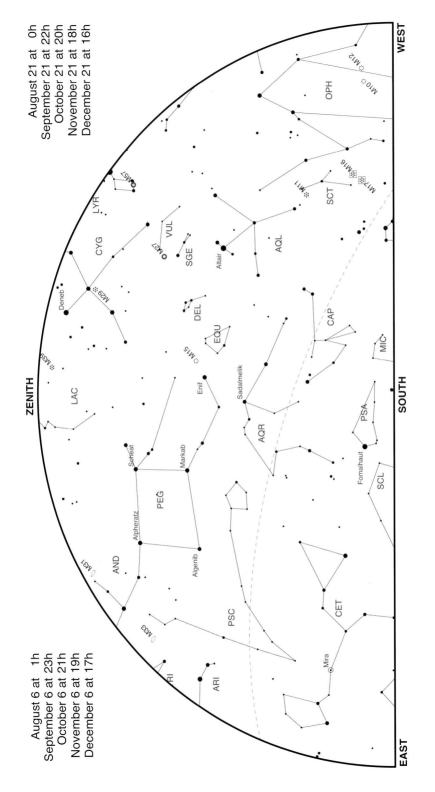

9S

WEST

August 21 at 0h
September 21 at 22h
October 21 at 20h
November 21 at 18h
December 21 at 16h

August 6 at 1h
September 6 at 23h
October 6 at 21h
November 6 at 19h
December 6 at 17h

ZENITH

SOUTH

EAST

10N

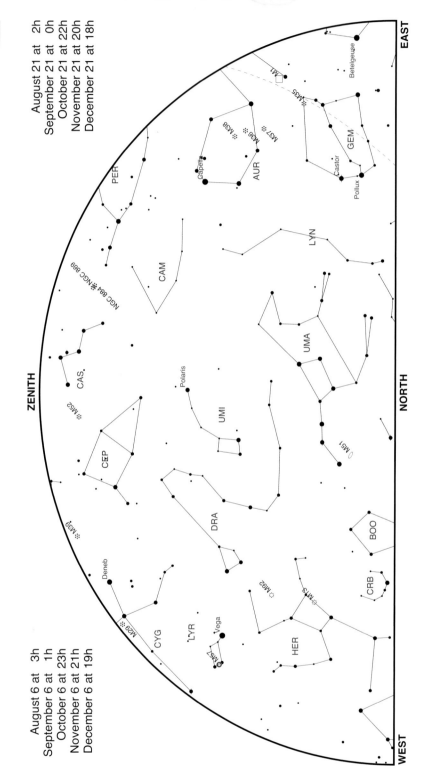

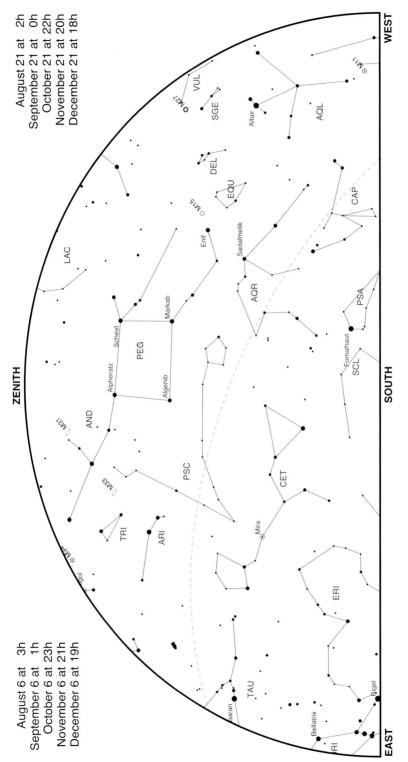

10S

WEST

August 21 at 2h
September 21 at 0h
October 21 at 22h
November 21 at 20h
December 21 at 18h

ZENITH

August 6 at 3h
September 6 at 1h
October 6 at 23h
November 6 at 21h
December 6 at 19h

EAST

SOUTH

VUL
M27
SGE
Altair
AQL
M11

DEL
EQU
M15
Enif
Sadalmelik
AQR
CAP
LAC
Scheat
Alpheratz
Markab
PEG
Algenib
PSA
Fomalhaut
SCL
AND
M31
M33
PSC
CET
Mira
TRI
ARI
ERI
Algol
M34
TAU
aran
Bellatrix
Rigel
RI

11N

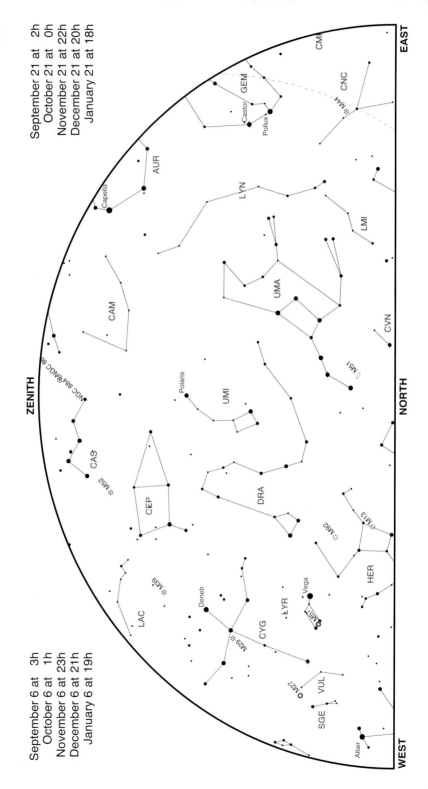

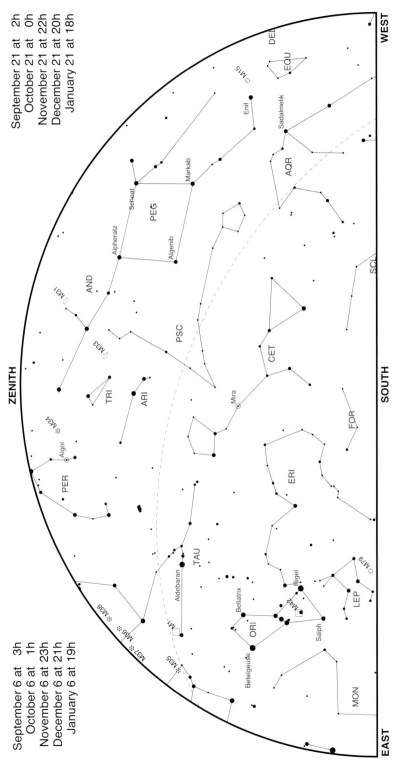

11S

WEST

ZENITH

EAST

SOUTH

September 21 at 2h
October 21 at 0h
November 21 at 22h
December 21 at 20h
January 21 at 18h

September 6 at 3h
October 6 at 1h
November 6 at 23h
December 6 at 21h
January 6 at 19h

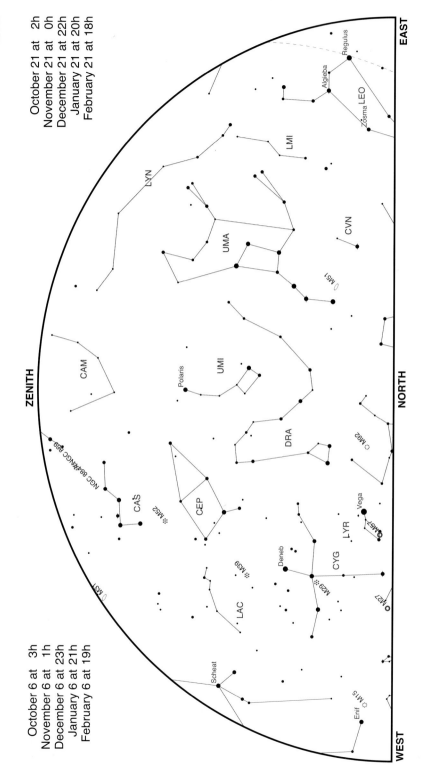

12N

October 21 at 2h
November 21 at 0h
December 21 at 22h
January 21 at 20h
February 21 at 18h

October 6 at 3h
November 6 at 1h
December 6 at 23h
January 6 at 21h
February 6 at 19h

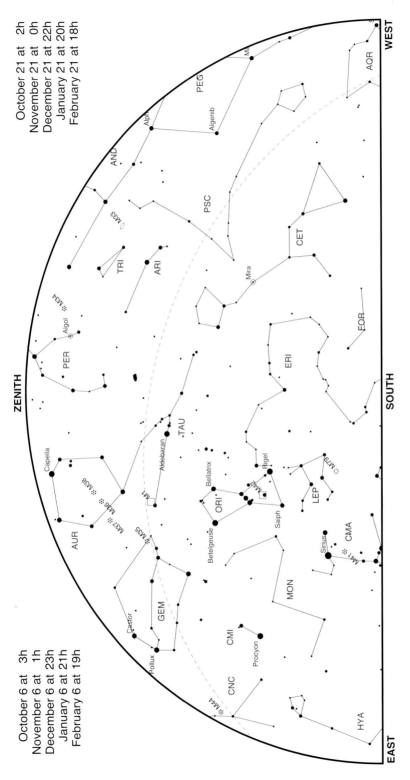

12S

WEST

EAST

SOUTH

ZENITH

October 21 at 2h
November 21 at 0h
December 21 at 22h
January 21 at 20h
February 21 at 18h

October 6 at 3h
November 6 at 1h
December 6 at 23h
January 6 at 21h
February 6 at 19h

Southern Hemisphere Star Charts

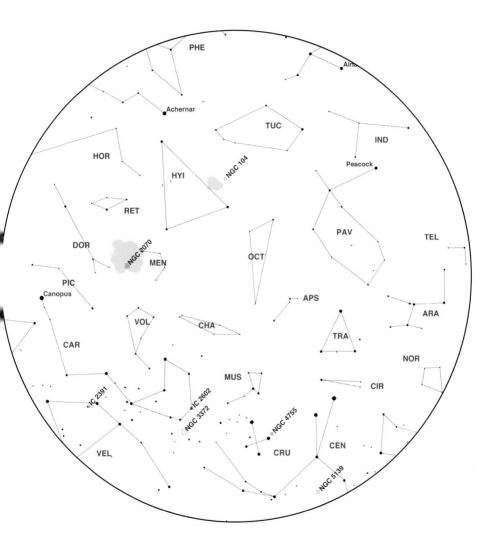

This chart shows stars lying at declinations between −45 and −90 degrees. These constellations are circumpolar for observers in Australia and New Zealand.

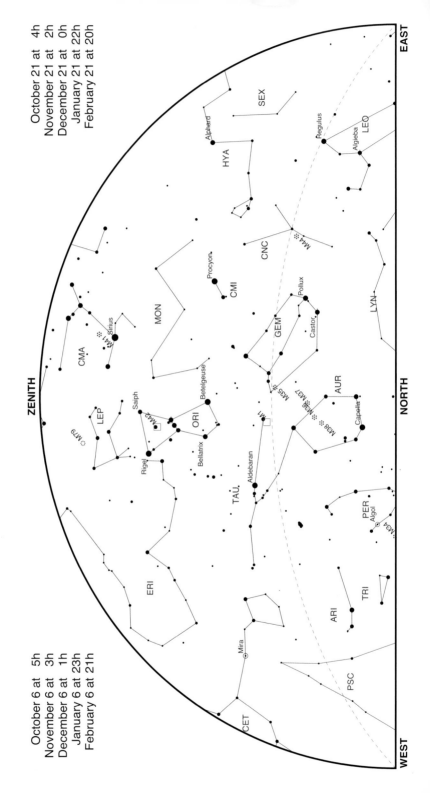

1N

October 21 at 4h
November 21 at 2h
December 21 at 0h
January 21 at 22h
February 21 at 20h

October 6 at 5h
November 6 at 3h
December 6 at 1h
January 6 at 23h
February 6 at 21h

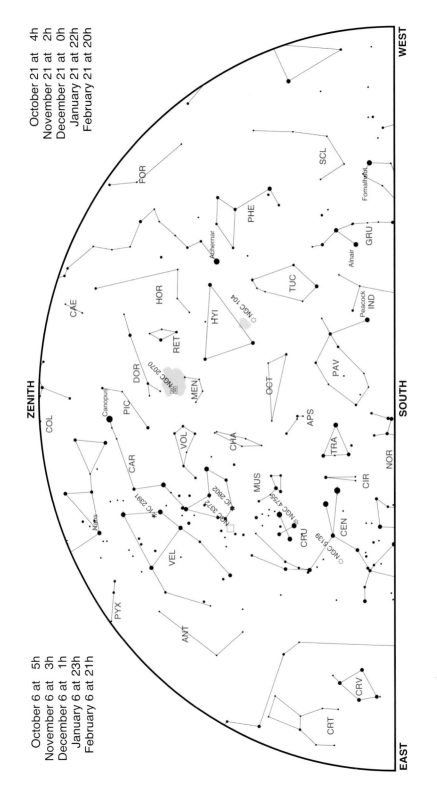

1S

WEST

EAST

ZENITH

SOUTH

October 21 at 4h
November 21 at 2h
December 21 at 0h
January 21 at 22h
February 21 at 20h

October 6 at 5h
November 6 at 3h
December 6 at 1h
January 6 at 23h
February 6 at 21h

FOR

SCL

PHE

Achernar

GRU

Alnair

Fomalhaut

TUC

NGC 104

HOR

HYI

CAE

RET

Peacock
IND

PAV

DOR

NGC 2070

PIC

MEN

OCT

COL

Canopus

CAR

VOL

CHA

APS

TRA

Mars

IC 2391

NGC 3372

IC 2602

MUS

NOR

CIR

PYX

VEL

NGC 4755

CRU

NGC 5139

CEN

ANT

CRV

CRT

2N

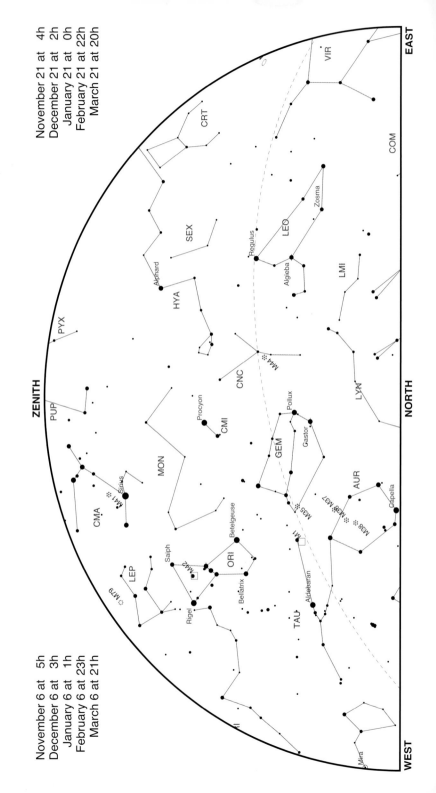

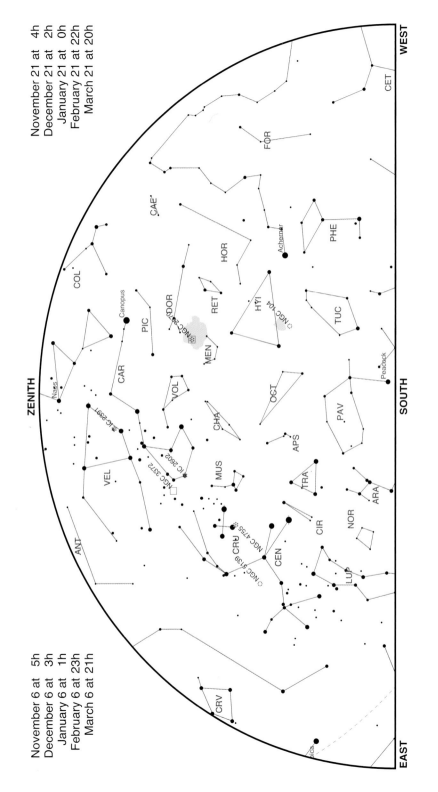

2S

November 21 at 4h
December 21 at 2h
January 21 at 0h
February 21 at 22h
March 21 at 20h

November 6 at 5h
December 6 at 3h
January 6 at 1h
February 6 at 23h
March 6 at 21h

WEST

ZENITH

SOUTH

EAST

CET
FOR
CAE
HOR
Achernar
PHE
NGC 2070
DORADO
RET
HYI
NGC 104
TUC
Peacock
COL
Canopus
PIC
MEN
OCT
PAV
CAR
VOL
CHA
APS
Naos
IC 2391
VEL
NGC 3372
IC 2602
MUS
TRA
ARA
ANT
NGC 4755
CRU
NGC 5139
CEN
CIR
NOR
LUP
CRV
Spica

3N

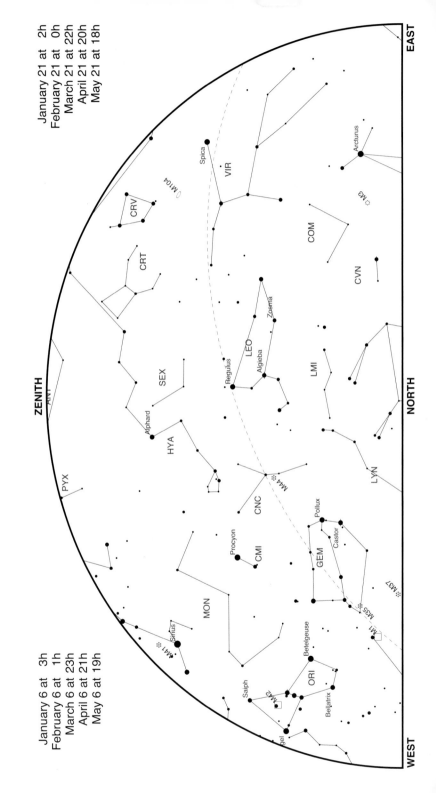

January 6 at 3h
February 6 at 1h
March 6 at 23h
April 6 at 21h
May 6 at 19h

EAST

ZENITH

NORTH

WEST

CRV

M104

Spica

VIR

Arcturus

M3

CRT

COM

CVN

SEX

LEO

Zosma

Regulus

Algieba

LMI

Alphard

HYA

PYX

LYN

CNC

M44

Procyon

CMI

GEM

Pollux

Castor

M37

M35

M1

MON

Betelgeuse

ORI

Saiph

M42

Bellatrix

Sirius

M41

Rigel

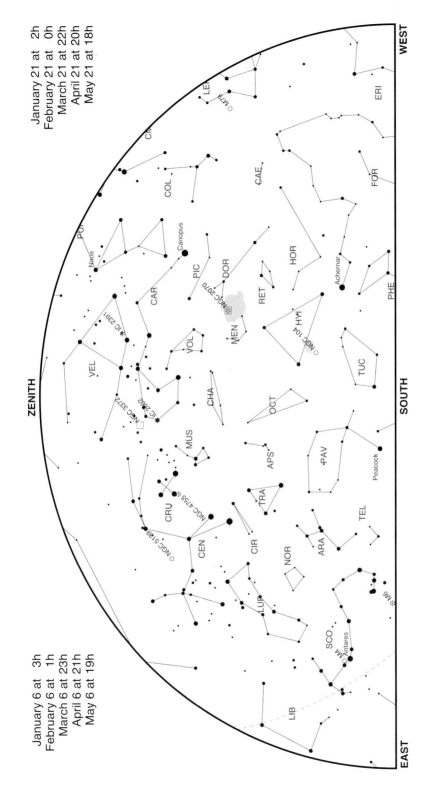

3S

WEST

January 21 at 2h
February 21 at 0h
March 21 at 22h
April 21 at 20h
May 21 at 18h

January 6 at 3h
February 6 at 1h
March 6 at 23h
April 6 at 21h
May 6 at 19h

ZENITH

SOUTH

EAST

4N

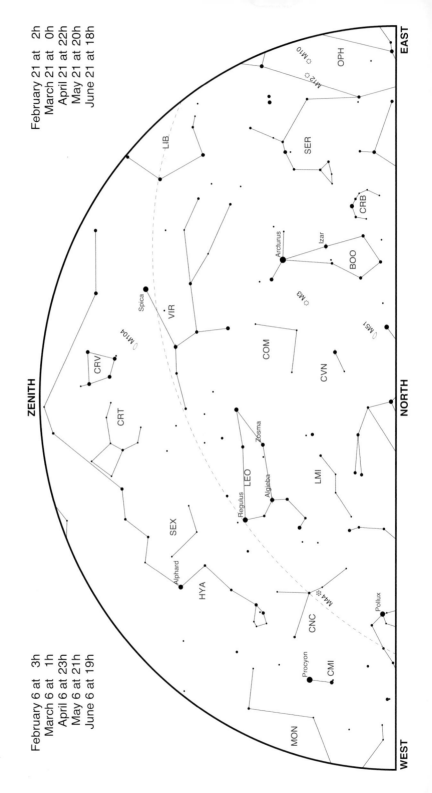

February 21 at 2h
March 21 at 0h
April 21 at 22h
May 21 at 20h
June 21 at 18h

February 6 at 3h
March 6 at 1h
April 6 at 23h
May 6 at 21h
June 6 at 19h

EAST

ZENITH

NORTH

WEST

OPH
M10
M12
SER
CRB
LIB
Arcturus
Izar
BOO
M3
Spica
VIR
M51
COM
CVN
CRV
M104
CRT
LMI
Zosma
LEO
Algieba
Regulus
SEX
Alphard
M44
CNC
HYA
Pollux
Procyon
CMI
MON

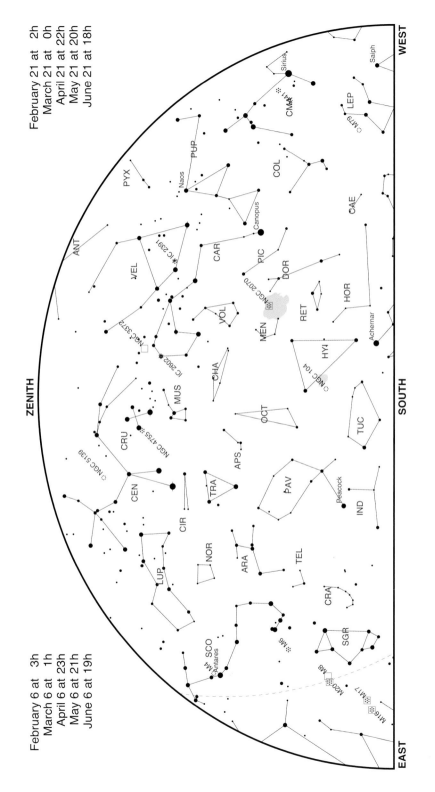

February 21 at 2h
March 21 at 0h
April 21 at 22h
May 21 at 20h
June 21 at 18h

February 6 at 3h
March 6 at 1h
April 6 at 23h
May 6 at 21h
June 6 at 19h

4S

WEST

ZENITH

EAST

SOUTH

5N

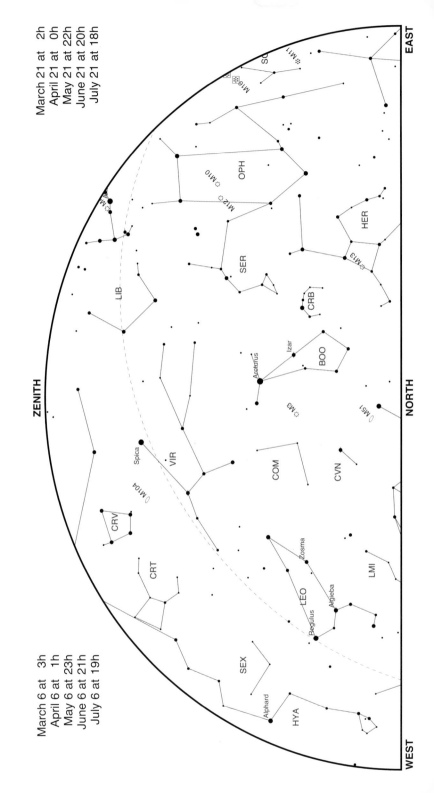

EAST

ZENITH

NORTH

WEST

March 6 at 3h
April 6 at 1h
May 6 at 23h
June 6 at 21h
July 6 at 19h

SCO
M11
M16
M17
OPH
M10
M12
HER
SER
M13
CRB
LIB
Izar
BOO
Arcturus
M3
M51
CVN
COM
Spica
VIR
M104
CRV
CRT
LEO
Zosma
Algieba
LMI
Regulus
SEX
Alphard
HYA

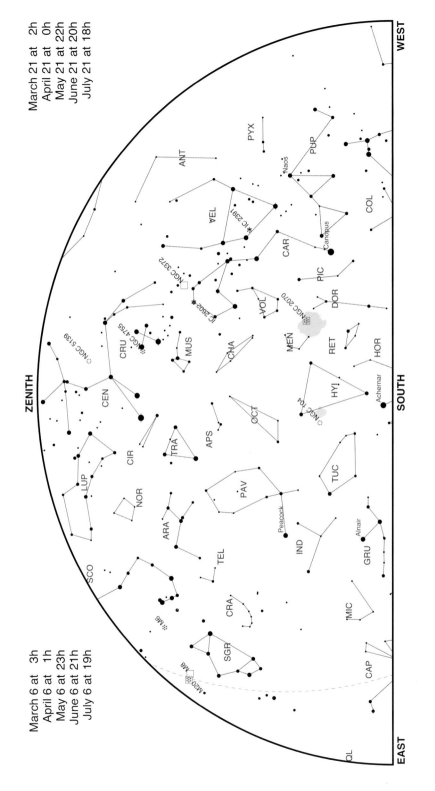

5S

March 21 at 2h
April 21 at 0h
May 21 at 22h
June 21 at 20h
July 21 at 18h

March 6 at 3h
April 6 at 1h
May 6 at 23h
June 6 at 21h
July 6 at 19h

WEST

EAST

SOUTH

ZENITH

PYX
ANT
VEL
*IC 2391
NGC 3372
CRU
NGC 4755
NGC 5139
CEN
IC 2602*
MUS
CHA
VOL
CAR
Canopus
Naos
PUP
COL
PIC
DOR
NGC 2070
MEN
RET
HOR
Achernar
NGC 104
HYI
TUC
OCT
APS
TRA
CIR
LUP
SCO
NOR
ARA
TEL
PAV
Peacock
IND
GRU
Alnair
MIC
CRA
SGR
M6
M20·M8
CAP

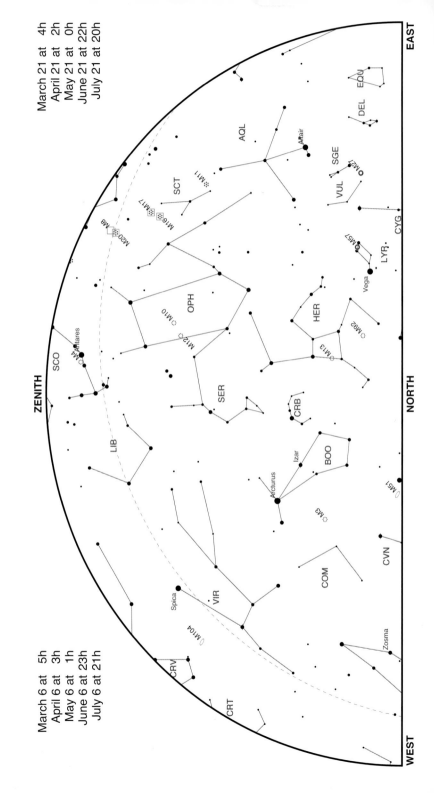

6N

March 21 at 4h
April 21 at 2h
May 21 at 0h
June 21 at 22h
July 21 at 20h

March 6 at 5h
April 6 at 3h
May 6 at 1h
June 6 at 23h
July 6 at 21h

EAST

WEST

NORTH

ZENITH

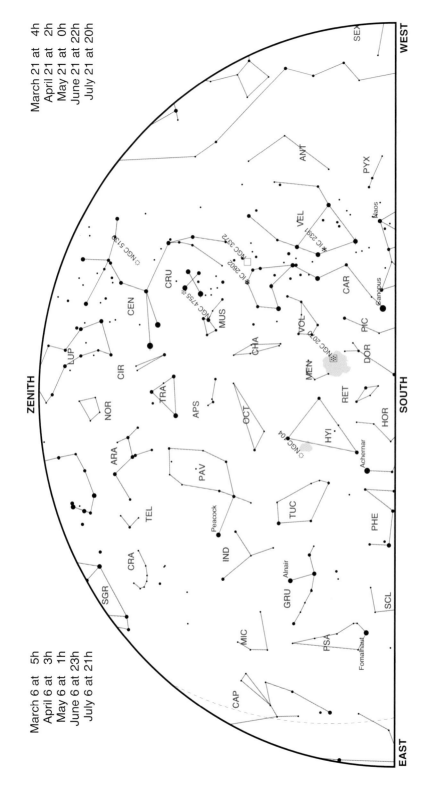

6S

WEST

March 21 at 4h
April 21 at 2h
May 21 at 0h
June 21 at 22h
July 21 at 20h

March 6 at 5h
April 6 at 3h
May 6 at 1h
June 6 at 23h
July 6 at 21h

ZENITH

EAST

SOUTH

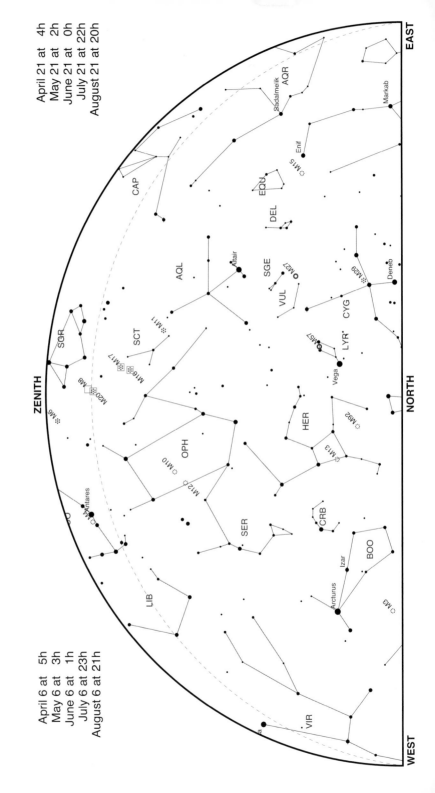

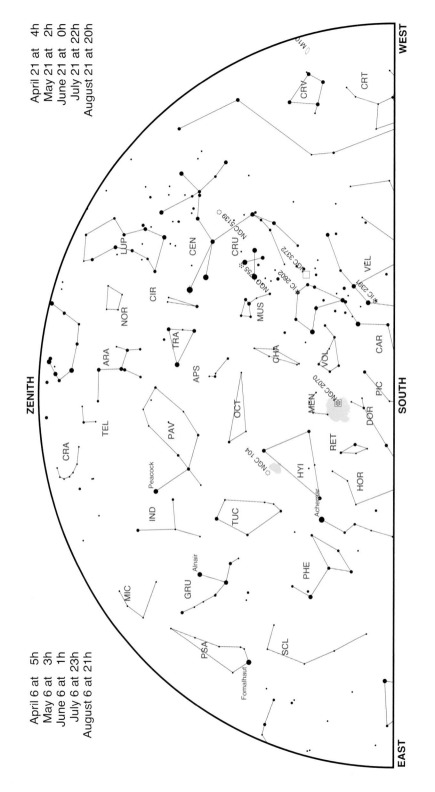

7S

WEST

ZENITH

EAST

SOUTH

CRV

CRT

LUP

CEN

CRU

NGC 5139

VEL

MGC 3372

IC 2391

CIR

NOR

IC 2602

NGC 755

MUS

CAR

TRA

APS

CHA

VOL

PIC

ARA

TEL

OCT

MEN

NGC 2070

DOR

CRA

PAV

RET

Peacock

HOR

IND

NGC 104

HYI

Achernar

TUC

MIC

GRU

Alnair

PHE

PSA

SCL

Fomalhaut

8N

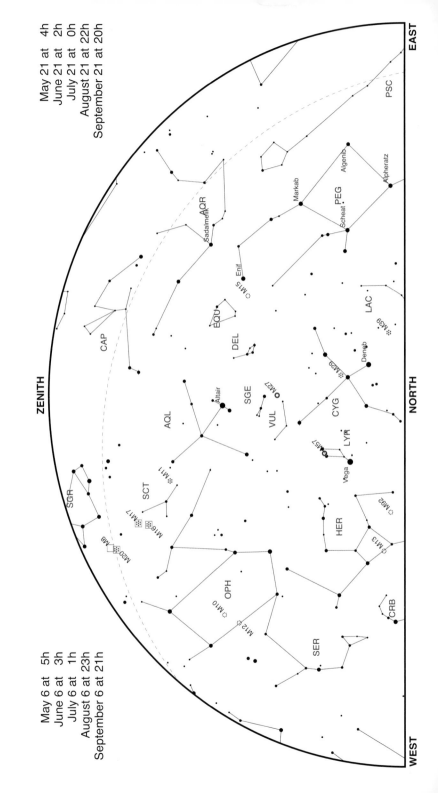

May 21 at 4h
June 21 at 2h
July 21 at 0h
August 21 at 22h
September 21 at 20h

May 6 at 5h
June 6 at 3h
July 6 at 1h
August 6 at 23h
September 6 at 21h

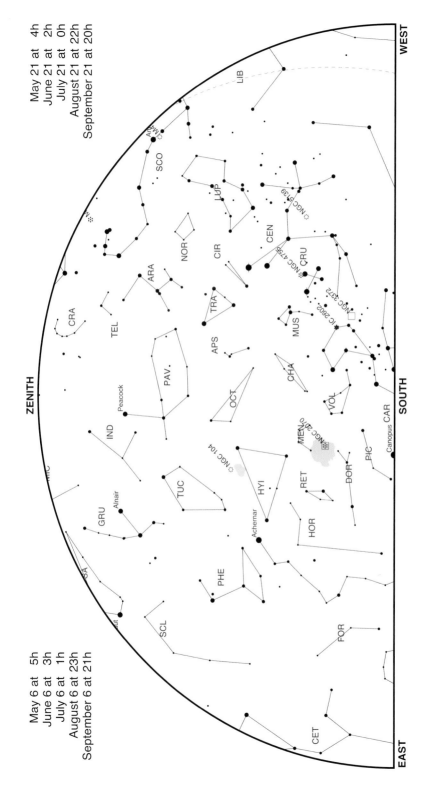

8S

WEST

May 21 at 4h
June 21 at 2h
July 21 at 0h
August 21 at 22h
September 21 at 20h

ZENITH

May 6 at 5h
June 6 at 3h
July 6 at 1h
August 6 at 23h
September 6 at 21h

EAST

SOUTH

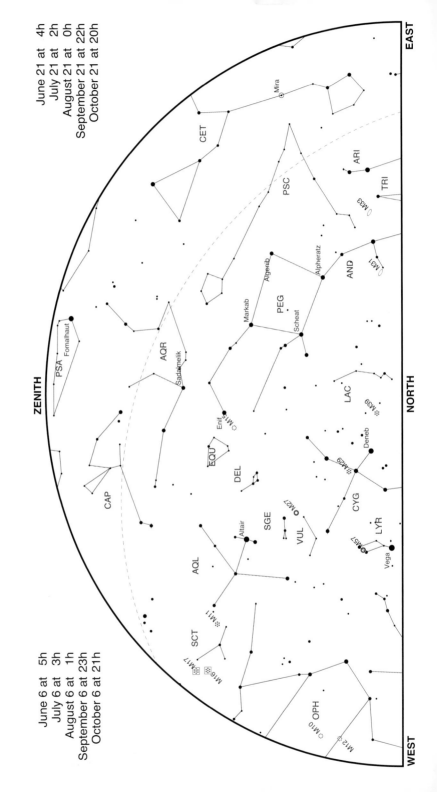

9N

June 21 at 4h
July 21 at 2h
August 21 at 0h
September 21 at 22h
October 21 at 20h

June 6 at 5h
July 6 at 3h
August 6 at 1h
September 6 at 23h
October 6 at 21h

EAST

ZENITH

NORTH

WEST

CET
ARI
TRI
M33
PSC
Algenib
Alpheratz
AND
M31
Markab
PEG
Scheat
Fomalhaut
PSA
AQR
Sadalmelik
LAC
M39
Enif
M15
Deneb
M29
EQU
DEL
CYG
CAP
LYR
M57
SGE
M27
VUL
Vega
Altair
AQL
M11
SCT
M17
M16
OPH
M10
M12
Mira

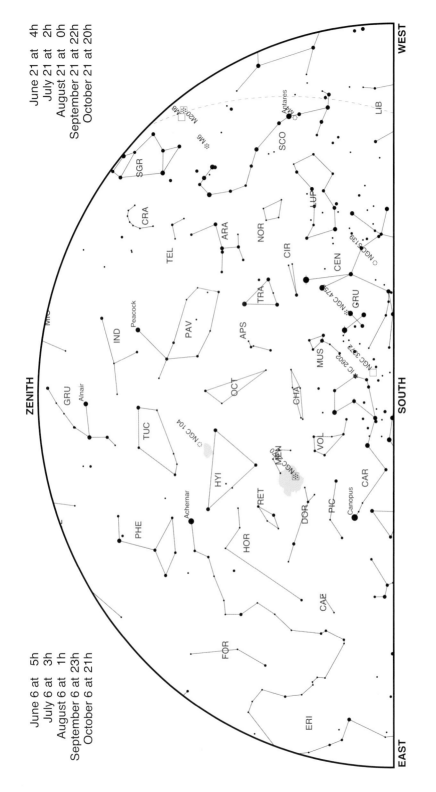

9S

June 21 at 4h
July 21 at 2h
August 21 at 0h
September 21 at 22h
October 21 at 20h

June 6 at 5h
July 6 at 3h
August 6 at 1h
September 6 at 23h
October 6 at 21h

WEST

ZENITH

SOUTH

EAST

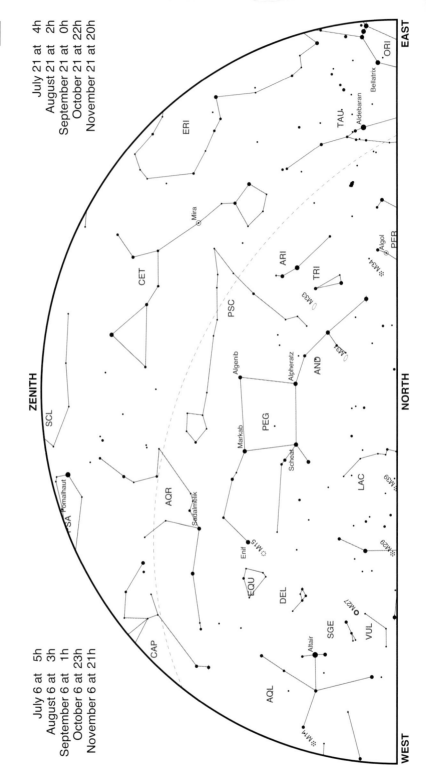

10N

July 6 at 5h
August 6 at 3h
September 6 at 1h
October 6 at 23h
November 6 at 21h

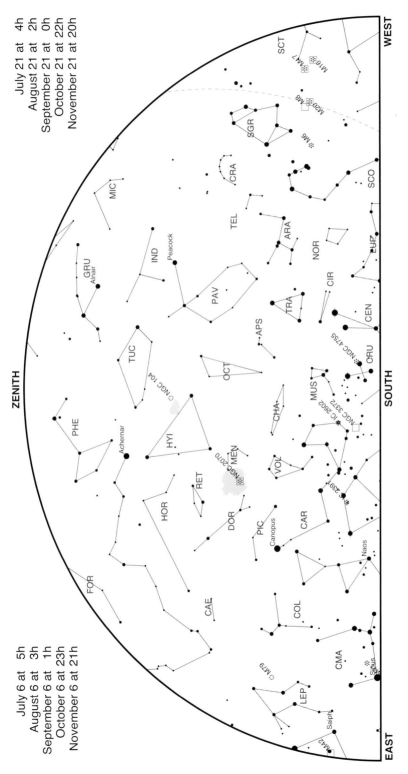

10S

WEST

EAST

ZENITH

SOUTH

July 21 at 4h
August 21 at 2h
September 21 at 0h
October 21 at 22h
November 21 at 20h

July 6 at 5h
August 6 at 3h
September 6 at 1h
October 6 at 23h
November 6 at 21h

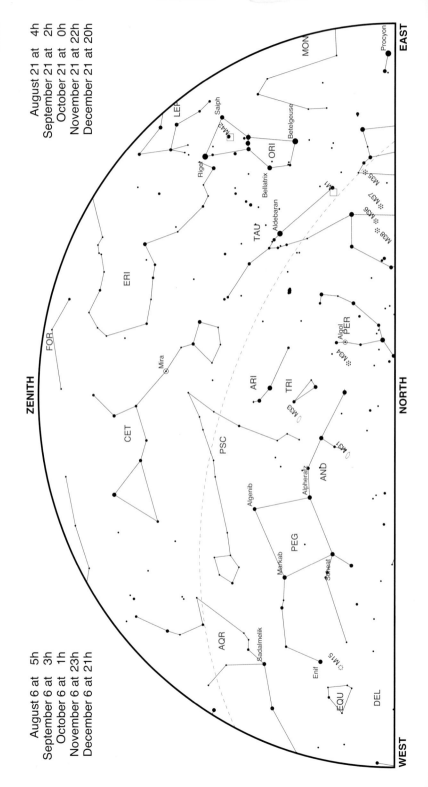

11N

August 21 at 4h
September 21 at 2h
October 21 at 0h
November 21 at 22h
December 21 at 20h

August 6 at 5h
September 6 at 3h
October 6 at 1h
November 6 at 23h
December 6 at 21h

ZENITH

EAST

WEST

NORTH

MON
LEP
Saiph
M42
ORI
Betelgeuse
Rigel
Bellatrix
TAU
Aldebaran
ERI
FOR
Mira
CET
ARI
TRI
M33
PSC
AND
M31
Alpheratz
Algenib
PEG
Markab
Scheat
AQR
Sadalmelik
Enif
M15
EQU
DEL
PER
Algol
M34
M35
M37
M36
M38
M1
Procyon

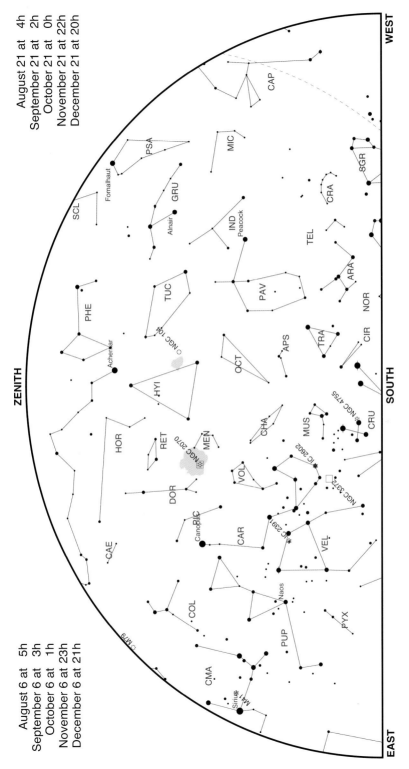

11S

WEST

August 21 at 4h
September 21 at 2h
October 21 at 0h
November 21 at 22h
December 21 at 20h

ZENITH

August 6 at 5h
September 6 at 3h
October 6 at 1h
November 6 at 23h
December 6 at 21h

EAST

SOUTH

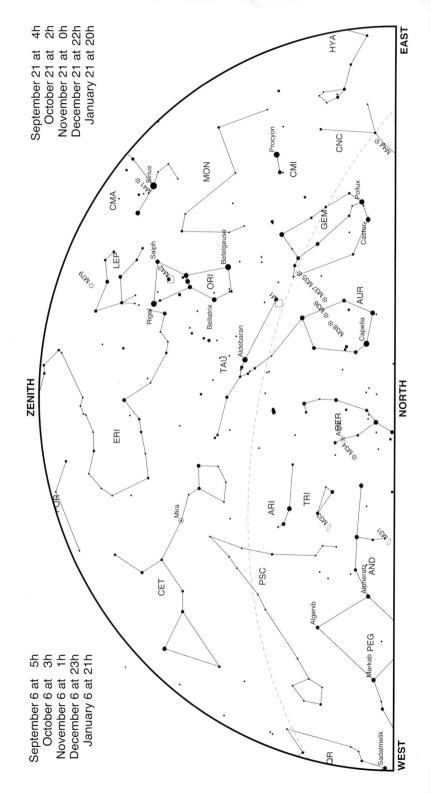

12N

September 21 at 4h
October 21 at 2h
November 21 at 0h
December 21 at 22h
January 21 at 20h

September 6 at 5h
October 6 at 3h
November 6 at 1h
December 6 at 23h
January 6 at 21h

ZENITH

EAST

NORTH

WEST

HYA

CNC
M44

Procyon
CMI

MON

Sirius
M41
CMA

Pollux
GEM
Castor

LEP
Saiph
M42
ORI
Betelgeuse
Bellatrix

M79

Rigel

M37 M35
M38 M36
AUR
Capella

M1

Aldebaran
TAU

AND
M34

ERI

ARI
TRI
M33

M31
AND
Alpheratz

Mira

Algenib

FOR

CET
PSC

Markab PEG

Sadalmelik
QR

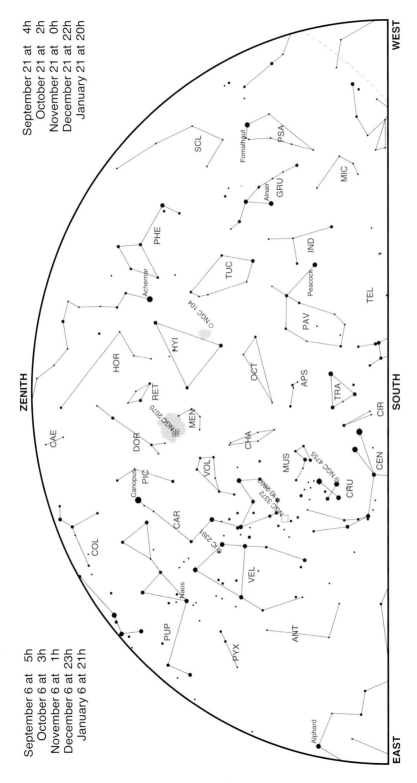

12S

WEST

EAST

SOUTH

ZENITH

September 21 at 4h
October 21 at 2h
November 21 at 0h
December 21 at 22h
January 21 at 20h

September 6 at 5h
October 6 at 3h
November 6 at 1h
December 6 at 23h
January 6 at 21h

The Planets in 2020

Lynne Marie Stockman

An unusual event takes place in late July, when all seven planets are (just) above the horizon simultaneously. On 24 July, an hour before sunrise, you can see Mercury (in Gemini), Venus (Taurus), Uranus (Aries), Mars (Pisces), Neptune (Aquarius), Saturn and Jupiter (both in Sagittarius). However, Mercury, Saturn and Jupiter are very close to the horizon, and a telescope will be necessary to spot faint Uranus and Neptune in the brightening sky. (The Moon is in the evening sky at this time and not part of this tableau).

Another notable occurrence happens on 21 December when Jupiter and Saturn are in conjunction for the first time in 20 years. First the two planets share a right ascension at around 13:30 UT and then five hours later they are at the same ecliptic longitude. Only 0.1° apart at closest approach, the gas giants are approximately 30° from the Sun and visible in the west after sunset. Jupiter is the brighter of the two objects, at magnitude –2.0, with Saturn considerably dimmer at +0.6. For more information relating to this event, see David Harper's article *Jupiter and Saturn – At Their Closest in Almost Four Centuries* elsewhere in this volume.

Mercury appears in both the morning and evenings skies four times this year. The best morning apparition for viewers in northern temperate latitudes is in October–December when the elusive planet reaches an altitude of nearly 20° at sunrise. These same observers will see Mercury at its best in the evening during the spring months of May and June. The situation for southern hemisphere planet watchers is somewhat different, with Mercury soaring nearly 30° above the eastern horizon at sunrise during the March–April apparition (when the planet reaches a greatest elongation west of 28°) and getting nearly as high above the western horizon from late August to late October. Mercury is less than half a degree north of first-magnitude star Spica on 22 September.

Venus is well-placed for viewing from northern latitudes this year, with fine apparitions in the evening from January to May, and in the morning from June to December. As an evening star, it is not as high in the sky for those in the southern hemisphere, slowly decreasing in altitude until May when it plummets to the horizon. It quickly reappears in the northeast in June, reaching a maximum altitude of over 30° at sunrise in late July and gradually sinking back toward the horizon for the rest of the year. The evening star passes very close to the open cluster M45 (the Pleiades) in early April and as the morning star, Venus is only 0.1° away from first-magnitude star Regulus on 2 October. It is occulted twice by the Moon, first in June (which may not be observable as the planet is at inferior conjunction just two weeks earlier) and again in December.

Mars is at opposition in October this year, reaching magnitude –2.6 in the constellation of Pisces. Earlier in the year, the red planet undergoes conjunctions in March with both Jupiter and Saturn, and five lunar occultations. It begins the year in Libra, over 2 AU from Earth and only second magnitude. It slowly gets closer and brighter, passing through Scorpius, Ophiuchus, Sagittarius, Capricornus, Aquarius, Pisces, and briefly Cetus before returning to Pisces in July where it spends the rest of 2020. A morning sky object in January, Mars rises earlier and earlier, passing into the evening sky in March for southern hemisphere observers, with northerners having to wait until July.

Jupiter begins 2020 in Sagittarius, moving to neighbouring Capricornus just before the end of the year. It is occulted by the Moon twice, in January and February, and has two particularly close conjunctions with other planets, namely Mars (in March) and Saturn (in December). Jupiter is a morning sky object at the beginning of the year, rising just before the Sun for observers in northern temperate latitudes and appearing in the east around 3am for those watching from the southern hemisphere. It rises a little earlier every morning, eventually appearing in the evening sky in March for southern latitudes and May for those farther north. The largest of the planets reaches opposition in July when it attains a magnitude of –2.8 but it does not undergo conjunction with the Sun in 2020. For the rest of the year Jupiter is an evening sky object, best seen from the southern hemisphere.

Saturn spends the year moving back and forth across the Sagittarius/ Capricornus border. It is at solar conjunction in mid-January, and emerges from the solar glare to become a morning sky object, rising about an hour before the Sun by mid-February at northern temperate latitudes. Like Jupiter, it has a conjunction with Mars in March and reaches opposition in July when it attains magnitude +0.2. The two gas giants gradually come together over the course of the year, meeting at conjunction in December. (This last happened in May 2000 and will not happen again until November 2040). The rings of the planet are gradually closing as the 2025 ring plane crossing approaches. They are open at an angle of 23.6° to the Earth at the beginning of the year before reaching a minimum of 20.5° in May, then opening again to 22.8° in October and closing slightly to end the year at 21.1°.

Uranus is located in Aries throughout 2020, visible in the evening sky at the beginning of the year but setting ever earlier on its way to solar conjunction in April. It reappears in the dawn sky in May, rising before midnight from July and reaching opposition in October when it is at its brightest (magnitude +5.7). It is visible in evening skies for the rest of the year. Uranus is not occulted by the Moon during 2020, nor does it pass near any bright stars. The planet can be glimpsed with the naked eye under perfect dark-sky conditions, but most observers will need binoculars or a small telescope.

Neptune is an eighth-magnitude object in the constellation Aquarius. It rises before Uranus, some two hours earlier when observed from northern latitudes and around four hours earlier as seen from the southern hemisphere. It begins the year in the evening sky, and passes near fourth-magnitude Phi (φ) Aquarii on 11 February on its way to solar conjunction in March. Neptune emerges into the morning sky before sunrise in April, rising earlier every night and attaining opposition in September where it reaches its brightest magnitude of +7.8. A small telescope is recommended for viewing this outermost planet of the solar system.

Phases of the Moon in 2020

Month	First Quarter	Full Moon	Last Quarter	New Moon
January	3	10	17	24
February	2	9	15	23
March	2	9	16	24
April	1 and 30	8	14	23
May	30	7	14	22
June	28	5	13	21
July	27	5	13	20
August	25	3	11	19
September	24	2	10	17
October	23	1 and 31	10	16
November	22	30	8	15
December	21	30	8	14

Eclipses in 2020

There are a minimum of four eclipses in any one calendar year, comprising two solar eclipses and two lunar eclipses. Most years have only four, although it is possible to have five, six or even seven eclipses during the course of a year, and in 2020 there are a total of six. This is close to the maximum, the last year in which there were seven eclipses being 1982, and the next 2038.

On 10 January there will be a penumbral lunar eclipse visible throughout most of Europe, Africa, Asia, the Indian Ocean and Western Australia. Maximum eclipse takes place at 19:10 UT, the eclipse beginning at 17:08 UT and ending at 21:12 UT.

There will be a penumbral lunar eclipse on 5 June visible throughout most of Europe, Africa, Asia, Australia, the Indian Ocean and Australia. The eclipse begins at 17:46 UT and ends at 21:04 UT with maximum eclipse occurring at 19:25 UT.

The path of the annular solar eclipse occurring on 21 June will begin in central Africa and travel through Saudi Arabia, northern India and southern China before ending in the Pacific Ocean. A partial eclipse will be visible throughout most of eastern Africa, the Middle East and southern Asia. The eclipse commences at 03:46 UT and ends at 09:34 UT with the full eclipse lasting from 04:48 UT to 08:32 UT and maximum eclipse at 06:40 UT.

Another penumbral lunar eclipse will take place on 5 July, beginning at 03:07 UT and ending at 05:52 UT. Maximum eclipse will be at 04:30 UT, the event being visible throughout most of North America, South America, the eastern Pacific Ocean, the western Atlantic Ocean and extreme western Africa.

The final penumbral lunar eclipse of the year will take place on 30 November and will be visible throughout most of North America, the Pacific Ocean and north eastern Asia including Japan. The eclipse begins at 07:32 UT and ends at 11:53 UT with maximum eclipse at 09:43 UT.

The path of totality of the total solar eclipse of 14 December will pass from the southern Pacific to the southern Atlantic, making landfall only in a narrow region of Chile and Argentina around 800 km south of Santiago and Buenos Aires. A partial eclipse will be visible in most parts of southern South America, the south eastern Pacific Ocean and the southern Atlantic Ocean. The eclipse begins at 13:34 UT and ends at 18:53 UT with total eclipse taking place between 14:32 UT and 17:54 UT and maximum eclipse occurring at 16:13 UT.

Some Events in 2020

January	3/4	Earth	Quadrantid Meteor Shower (ZHR 120)
	5	Earth	Perihelion (0.9832 AU)
	10	Mercury	Superior Conjunction
	10	Earth/Moon	Penumbral Lunar Eclipse
	13	Saturn	Conjunction
	23	Jupiter	Lunar Occultation
	23	Uranus	East Quadrature
	27	Venus, Neptune	Conjunction (0.1° apart, 40° from Sun)

February	10	Mercury	Greatest Elongation East (18°)
	18	Mars	Lunar Occultation
	19	Jupiter	Lunar Occultation
	26	Mercury	Inferior Conjunction

March	8	Neptune	Conjunction
	18	Mars	Lunar Occultation
	20	Earth	Equinox
	20	Mars, Jupiter	Conjunction (0.7° apart, 68° from Sun)
	24	Mercury	Greatest Elongation West (28°)
	24	Moon	Farthest Apogee of the Year
	24	Venus	Greatest Elongation East (46°)
	31	Mars, Saturn	Conjunction (0.9° apart, 71° from Sun)

April	2	3 Juno	Opposition in Virgo (magnitude +9.5)
	3	Venus	0.3° South of the Pleiades (M45)
	7	Moon	Nearest Perigee of the Year
	8	Moon	Super Moon
	15	Jupiter	West Quadrature
	21	Saturn	West Quadrature
	22/23	Earth	Lyrid Meteor Shower (ZHR 18)
	26	Uranus	Conjunction

May	4	Mercury	Superior Conjunction
	6/7	Earth	Eta (η) Aquarid Meteor Shower (ZHR 30)
	22	Mercury, Venus	Conjunction (0.9° apart, 19° from Sun)

	3	Venus	Inferior Conjunction
	4	Mercury	Greatest Elongation East (24°)
	5	Earth/Moon	Penumbral Lunar Eclipse
	6	Mars	West Quadrature
June	11	Neptune	West Quadrature
	19	Venus	Lunar Occultation
	20	Earth	Solstice
	21	Earth/Moon	Annular Solar Eclipse

	1	Mercury	Inferior Conjunction
	4	Earth	Aphelion (1.017 AU)
	5	Earth/Moon	Penumbral Lunar Eclipse
	11	Venus	1.0° North of Aldebaran (α Tauri)
July	13	2 Pallas	Opposition in Sagitta (magnitude +9.6)
	14	Jupiter	Opposition in Sagittarius (magnitude −2.8)
	15	134340 Pluto	Opposition in Sagittarius (magnitude +14.5)
	20	Saturn	Opposition in Sagittarius (magnitude +0.2)
	22	Mercury	Greatest Elongation West (20°)
	28/29	Earth	Delta (δ) Aquarid Meteor Shower (ZHR 20)

	2	Uranus	West Quadrature
	9	Mars	Lunar Occultation
August	12/13	Earth	Perseid Meteor Shower (ZHR 80)
	13	Venus	Greatest Elongation West (46°)
	17	Mercury	Superior Conjunction
	28	1 Ceres	Opposition in Aquarius (magnitude +7.7)

	6	Mars	Lunar Occultation
	11	Neptune	Opposition in Aquarius (magnitude +7.8)
September	22	Mercury	0.3° North of Spica (α Virginis)
	22	Earth	Equinox

	1	Mercury	Greatest Elongation East (26°)
	1	Moon	Harvest Moon
	2	Venus	0.1° South of Regulus (α Leonis)
	3	Mars	Lunar Occultation
	7/8	Earth	Draconid Meteor Shower (ZHR 10)
	10	Earth	Southern Taurid Meteor Shower (ZHR 5)
October	11	Jupiter	East Quadrature
	13	Mars	Opposition in Pisces (magnitude −2.6)
	18	Saturn	East Quadrature
	21/22	Earth	Orionid Meteor Shower (ZHR 25)
	25	Mercury	Inferior Conjunction
	31	Moon	Blue Moon, Hunter's Moon, Micro Moon
	31	Uranus	Opposition in Aries (magnitude +5.7)

November	10	Mercury	Greatest Elongation West (19°)
	12	Earth	Northern Taurid Meteor Shower (ZHR 5)
	17/18	Earth	Leonid Meteor Shower (ZHR varies)
	30	Earth/Moon	Penumbral Lunar Eclipse

December	9	Neptune	East Quadrature
	12	Venus	Lunar Occultation
	13/14	Earth	Geminid Meteor Shower (ZHR 75+)
	14	Earth/Moon	Total Solar Eclipse
	20	Mercury	Superior Conjunction
	21	Earth	Solstice
	21	Jupiter, Saturn	Conjunction (0.1° apart, 30° from Sun)
	22	Earth	Ursid Meteor Shower (ZHR 10)

The entries for meteor showers state the date of peak shower activity (maximum). The figure quoted in parentheses in column 4 alongside each meteor shower entry is the expected Zenith Hourly Rate (ZHR) for that particular shower at maximum. For a more detailed explanation of ZHR, and for further details of the individual meteor showers listed here, please refer to the article *Meteor Showers in 2020* located elsewhere in this volume. For more on each of the eclipses occurring during the year, please refer to the information given in *Eclipses in 2020*.

Monthly Sky Notes and Articles

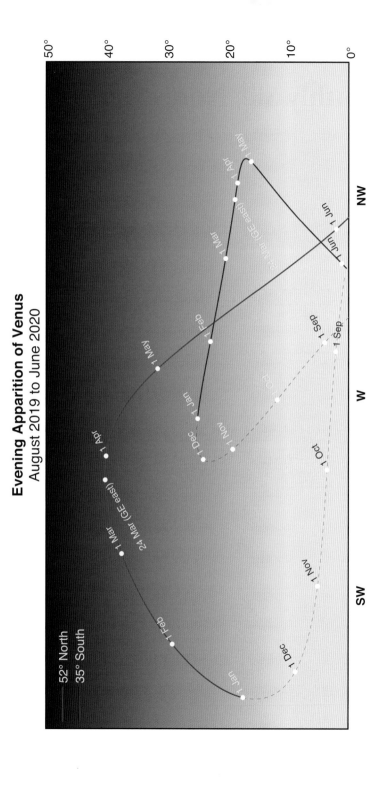

Evening Apparition of Venus
August 2019 to June 2020

52° North
35° South

January

Full Moon: 10 January
New Moon: 24 January

MERCURY arrives at superior conjunction on 10 January and is unobservable at the beginning of the year. Unfortunately, this means its conjunctions with Jupiter (2 January) and Saturn (12 January) will go unnoticed. Mercury soon reappears in the west, setting just after the Sun, in what is a decent evening apparition for observers in northern temperate latitudes. It is fairly bright too, shining at magnitude -1.0.

VENUS is the evening star for the first five months of 2020. This is an excellent apparition for observers in northern latitudes, with the bright planet eventually rising $40°$ or more above the western horizon at sunset. However, it begins the year low in the southwest. For those watching the planet from the southern hemisphere, Venus is never higher in the sky than at the beginning of the year. On 27 January, Venus is just $0.1°$ south of Neptune. Shining at magnitude -4.1, Venus is easily visible but a small telescope will be necessary to see eighth-magnitude Neptune nearby.

EARTH is at perihelion on 5 January. The Sun is only 0.9832 AU (147,100,000 km) away from our planet on this date. Five days later, the Full Moon undergoes a penumbral eclipse. Please refer to the section entitled *Eclipses in 2020* for more information.

MARS was at conjunction last September so the red planet is a morning sky object at the outset of 2020, rising about an hour before morning twilight. The first half of January is spent in the constellation of Scorpius, after which Mars moves into the non-zodiacal constellation of Ophiuchus. The waning crescent Moon moves past on 20 January. Mars begins the year over 2 AU away and at only $+1.6$ magnitude, brightening to $+1.4$ at the end of the month.

JUPITER was at conjunction at the end of December so its close approaches to third-magnitude star Lambda (λ) Sagittarii (Kaus Borealis), planet Mercury and fifth-magnitude globular cluster M22 will not be visible in the first week and a half of January. However, toward the end of the month, Jupiter will emerge from the solar glare to appear in the east before sunrise. A lunar occultation, visible from Madagascar, occurs on 23 January from around 01:30 UT.

SATURN is at conjunction on 13 January and thus is not visible this month. Its close encounters with Mercury on 12 January and the New Moon on 24 January will go unseen.

URANUS is already above the horizon at sunset in January. Slowly looping through the constellation of Aries this year, the planet reaches its maximum declination south (+11.8°) for 2020 on 9 January. Two days later, Uranus resumes direct motion after entering the year in retrograde. This sixth-magnitude planet arrives at east quadrature on 23 January. *A finder chart showing the position of Uranus throughout 2020 can be found in the October sky notes. Background stars are shown to magnitude 8.*

NEPTUNE continues its sojourn through the constellation of Aquarius. It is found in the evening sky, setting two hours before midnight at the beginning of January and disappearing two hours earlier than that by the end of the month. On 27 January, it is only 0.1° north of Venus but a telescope will be necessary to see the blue ice giant (magnitude +7.9). *A finder chart showing the position of Neptune throughout 2020 can be found in the September sky notes. Background stars are shown to magnitude 10.*

Dark Sky Places: Now and in the Future

Bob Mizon

Loss of the Night

Not very long ago, the whole world was a dark-sky place. Then, in the late nineteenth century, the rise of public outdoor lighting had its small beginnings in the work of English physicist and chemist Sir Joseph Wilson Swan and American physicist and inventor Thomas Alva Edison. Working independently, they unwittingly started a process that would spell the end of humanity's natural nights across much of the developed world by almost simultaneously perfecting sealed, bright incandescent-filament lamps, powered by electricity. By the mid-twentieth century, people in urban areas and their surroundings were no longer able to behold Nature's grandest free show: the spectacle of the untainted starry heavens. Exterior electric lamps had proliferated. Most were not designed to send emissions only where needed, but outwards and upwards too. So, by about 1960, the lights of the world had not only increased visibility on roads and pavements and in public spaces, but had created the night-time sky glow which has taken away the stars.

Facing Up to the Problem

In 1988, in the USA, the International Dark-Sky Association (IDA) began its work. The principal founder members were Dr David Crawford and Professor Tim Hunter, and the organisation's aim was to work against sky glow by promoting the kind of carefully designed exterior lighting that lit only its intended target, at an appropriate level of brightness. The IDA now has several thousand members in its branches in many countries. Not long afterwards, in 1989, a group of members of the British Astronomical Association (BAA), the UK's largest body representing the interests of all who wish to observe and enjoy the wonders of the night sky, founded its Committee (now Commission) for Dark Skies (CfDS). Their aim was to halt the tide of ill-designed lighting and to promote 'star-quality' types which put light where needed, in order to bring back the stars. Similar movements now exist all around the world.

The summer Milky Way over the UK's first Dark Sky Park (2009), Galloway Forest in Scotland. (James Hilder)

The IDA set up its International Dark Sky Places (IDSP) Program in 2001. Its aim is to "encourage communities, parks and protected areas around the world to preserve and protect dark sites through responsible lighting polices and public education". Areas which have committed themselves to the protection of their starry skies can now be found in every continent of the world, except Antarctica, where Nature itself guarantees a good view of the cosmos. They range in size from Mont-Mégantic, the world's first International Dark Sky Reserve, in Québec, Canada (5,300 square kilometres), to the Dark Sky Community of the island of Coll in Scotland (77 square kilometres). Dark Sky Reserves are usually found in extensive areas such as National Parks, with existing pristine night skies across much of their territory. Visitors are actively encouraged to experience designated dark sky sites, information programmes are established, stargazing events are organised and publicity distributed to local tourist offices, hotels, campsites and, if present, observatories. An International Dark Sky Reserve must specify a core area satisfying IDA criteria for natural darkness and night sky quality, with a peripheral buffer zone in support. Dark Sky Parks, usually smaller than the Reserves, are large-scale open spaces or publicly accessible areas of comparable size with exceptional vistas of the night sky. Organisers promote best use of their dark sky resource by monitoring both local and distant lighting, to facilitate star watching and benefit wildlife and the nocturnal environment, and they pursue educational and cultural programmes. Smaller still, Dark Sky Communities may be villages, towns, islands or other limited areas dedicated to the protection of the night sky, having taken measures to ensure best-quality lighting and effective public education about dark skies. There are three tiers (Gold, Silver, Bronze) within which IDA dark sky status is awarded. These represent the levels of visibility of the faintest stars seen from the IDSP areas, indicating the sky darkness according to the Bortle Classification (see link below).

LEDs and Starry Skies

Over the last decade, astronomers, environmentalists and health-related organisations such as the American Medical Association have raised concerns about the remarkably rapid shift in public lighting towards light-emitting diode (LED) technology. Its negative impacts on the night sky, the terrestrial environment and living things have been widely reported. In 2013 in Pamplona,

A typical modern LED domestic/commercial floodlight. Much of its emissions go into the sky. This one is in a National Park. (Bob Mizon)

Spain, delegates at the Thirteenth European Symposium for the Protection of the Night Sky issued a statement on European Union policy initiatives for accelerating the installation of LED lighting. They expressed their disquiet about what American astronomer Bob King called the 'tsunami of LED lights' that were rapidly replacing older sodium types. All artificial lighting at night has some impact on the environment and biodiversity, altering for example foraging patterns in wildlife and circadian cycles of living things in general (including humans). This is now especially true of the kind of blue-rich light which is much in evidence in the LED units favoured by many administrations and in private schemes. As far as the visibility of the night sky is concerned, blue light is readily scattered in the atmosphere and therefore leads to increased sky glow. If lights are well directed and not too bright for the lighting task so that upward reflections from surfaces are minimal, then LEDs can be the answer astronomers have been seeking to 'night blight'. Sadly, such good practice is too often absent. The former orange glow from sodium lamps is being replaced by the silvery-grey sheen from LEDs. The hopes that dark-sky campaigners had of LEDs are not being realised. These versatile, solid-state lights, some of which have earned their inventors a Nobel Prize, are too often used without bringing their best qualities into play. Those installing them seem not to know that they can be so precisely directed, are readily available in many colours and spectral registers, are dimmable, and can be remotely controlled in response to traffic fluctuations and other localised conditions.

Can Dark Sky Places Survive?

Officials in Dark Sky Reserves, Parks and Communities are finding their successful past and present efforts to control local lighting compromised by a new influx of light projected from new and distant LED sources outside their

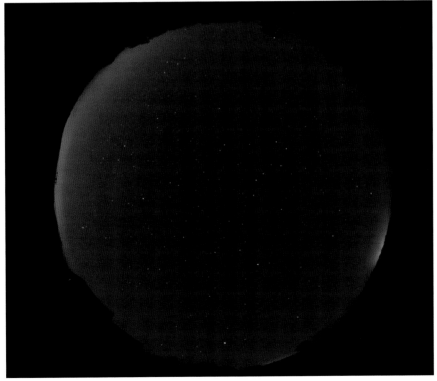

An all-sky photo from the Malvern Hills Area of Oustanding Natural Beauty. To the right, the skyglow from Hereford showing LED spill light replacing sodium as the town is re-lit. (Dr Chris Baddiley)

borders. Even at very shallow upward angles, stray light travels a very long way. Low-angle upward light meets denser concentrations of atmospheric particles and aerosols, increasing downward scatter. Continual growth of less-than-perfect lighting schemes near dark-sky sites may eventually overwhelm their stars, unless all who care about the night sky combine their voices to insist that more thought is given to lighting choices and design. It is of interest that the European Union, for example, incorporates the precautionary principle into its legal structure, but apparently not in the case of new LED lighting technology. The EU's definition of the Precautionary Principle is: "When human activities may lead to morally unacceptable harm that is scientifically plausible but uncertain, actions shall be taken to avoid or diminish that harm". Is it therefore against the spirit of such law, or of usual scientific practice, to promote bright blue-rich ('white') light sources at night despite the growing number of studies pointing out their deleterious effects? A moratorium on the further installation of blue-rich LEDs should be introduced to allow time for greater consideration of negative impacts – health, ecological, strategic, financial and of course celestial. Our approach to the issue of climate change increasingly stresses green energy sources, but we also need to stem the growth of energy requirements: one way to achieve this is by using only the right amount of light, when and where needed. Sustainable green energy plans make no sense while we are simultaneously allowed to waste energy. In an era when energy considerations loom ever larger, the promotion and installation of polluting lights seems a bad joke.

The light from the rest of the Universe takes hundreds, thousands, millions, even billions of years to reach our eyes. What a pity to lose it in the last millisecond of its enormous journey.

Useful References

International Dark-Sky Association: **www.darksky.org**
Lists of IDA Dark Sky Places worldwide: **www.darksky.org/idsp**
BAA Commission for Dark Skies: **www.britastro.org/dark-skies**
Bortle Classification: **en.wikipedia.org/wiki/Bortle_scale**
The Precautionary Principle: **www.precautionaryprinciple.eu**
Lighting and Circadian Rhythms: **www.energy.gov/sites/prod/files/2014/11/ f19/lockley_spectrum_detroit2014.pdf**

February

Full Moon: 9 February
New Moon: 23 February

MERCURY is visible in the evening this month and is best seen from the tropics although it reaches an altitude of around 15° for observers in northern temperate latitudes. The tiny planet, shining at magnitude −0.6, is found 18° away from the Sun at greatest elongation east on 10 February. Two days later Mercury arrives at the first of four perihelia this year. Retrograde motion begins on 16 February and ten days later, Mercury reaches inferior conjunction. Mercury's brightness varies greatly this month, beginning at magnitude −1.0 before dimming to sixth magnitude around time of conjunction. Mercury reappears at the end of the month in the east before sunrise.

Evening Apparition of Mercury
10 January to 25 February

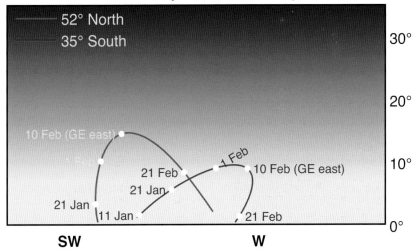

VENUS slowly gains altitude in the southwest as seen from the northern hemisphere. However, this is a much poorer apparition for viewers in southern latitudes, with the evening star slowly sinking back toward the horizon between January and May. A telescope reveals the planet to be in its waning gibbous phase and slowly brightening from magnitude −4.1 to −4.2 by the end of the month.

MARS reaches its descending node on the first day of the month, moving to a position south of the ecliptic. It leaves Ophiuchus for Sagittarius on 11 February and one week later, is occulted by the waning crescent Moon in an event that is visible from North America from around 11:30 UT. On 21 February, Mars reaches it most southerly declination, −23.7°, of the year. Mars is best viewed from the southern hemisphere where it rises before midnight. Northerly planet watchers get an hour or less of observing the red planet before dawn arrives.

JUPITER spends the month of February moving past the Teaspoon asterism of the constellation Sagittarius, passing a degree south of fourth-magnitude star Omicron (o) Sagittarii around 7 February. A morning sky object shining at magnitude −1.9, Jupiter is best seen from the southern hemisphere where it rises very early in the morning. Observers in northern temperate latitudes find the bright planet still embedded in morning twilight. Another lunar occultation occurs this month on 19 February, but dress warmly as you will need to be in the Antarctic to see it. Jupiter crosses the ecliptic north to south on 27 February but because of the tilt of the ecliptic with respect to the celestial equator, the planet's declination is actually increasing.

SATURN reappears in the dawn sky following last month's conjunction. Currently in the constellation of Sagittarius, the ringed planet crosses the ecliptic north to south on 15 February. Five days later Saturn is found 1.7° north of the waning crescent Moon. As the tilt of the rings lessens this month, Saturn fades slightly, from magnitude +0.6 to +0.7.

URANUS is an evening sky object in the faint constellation of Aries, best seen early in the evening from northern latitudes. At magnitude +5.8, Uranus requires dark skies and probably at least binoculars to view it.

NEPTUNE glides past fourth-magnitude Phi (φ) Aquarii on 11 February but with the planet only magnitude +7.9, a telescope is required to see this pairing during the early evening hours. Faint Neptune is soon lost in the evening twilight as it approaches conjunction next month.

Ashes to Ashes, Stardust to Stardust

Peter Rea

Only taxation lasts forever, but we mortals must one day shuffle off this mortal coil and then what happens to our remains? For some folks it means a trip to space, infinity and beyond!

When Clyde Tombaugh discovered Pluto on 18 February 1930 he could not have imagined that one day part of his ashes would fly past that celestial body (although by the time of that flyby, Pluto had been demoted to a dwarf planet). Clyde Tombaugh was 24 at the time of the discovery and died on 17 January 1997 aged 90. When the New Horizons spacecraft was launched toward Pluto on 19 January 2006 a small portion of Clyde Tombaugh's ashes were sealed in a small canister and placed aboard the spacecraft. On 14 July 2015 those ashes along with New Horizons flew past the "planet" he discovered. On 1 January 2019 that special canister flew past Kuiper Belt Object 2014 MU69, known by the name Ultima Thule, a fitting tribute to a great and patient observer. More details of the New Horizons encounter with Ultima Thule are given in the article *Solar System Exploration in 2019* elsewhere in this volume.

The year 2019 saw the 50th anniversary of the first Moon landing. The ability of the astronauts to perform geological descriptions and sampling was down to the training of many geologists from the United States Geological Survey (USGS) based at Menlo Park in California. One that stands out is Eugene Merle (Gene) Shoemaker who joined USGS early in his career. Shoemaker worked on the Ranger and Surveyor unmanned missions of the early 1960s. He became a Principal

Clyde Tombaugh is pictured here in 1930 standing alongside his homemade telescope. (Popular Science Monthly / Wikimedia Commons)

Investigator for the early Apollo missions and did much research on crater formation. Shoemaker died on 18 July 1997 aged 69. It seems fitting therefore that when the Lunar Prospector mission was launched to the Moon on 7 January 1998, a part of Shoemaker's ashes should be placed aboard the spacecraft. After a successful mission in lunar orbit the spacecraft was deliberately crashed onto the lunar surface on 31 July 1999. A part of Eugene Shoemaker had arrived on the Moon, a body he had studied for much of his illustrious career.

USGS photo of Eugene Merl (Gene) Shoemaker. (US Geological Survey / Wikimedia Commons)

Canadian actor James Doohan, known for playing Montgomery (Scotty) Scott, Chief Engineer on the Starship Enterprise had a portion of his ashes sent into Earth orbit but not at the first attempt. Doohan died on 20 July 2005 aged 85. A portion of his ashes were launched on a sub orbital rocket in 2007 and briefly entered "space" before falling back to Earth. The ashes were launched again the following year on an early Falcon 1 rocket that suffered a launch failure. It was a case of third time lucky for "Scotty" when some of Doohan's ashes entered Earth orbit in 2012 using a Falcon 9 rocket. It might have been easier to "beam" the ashes into space!

Sending ashes into space is now a commercial venture and many remains have already made it into space. An Internet search will reveal the companies offering this service.

For some time now, the American space agency NASA has been offering the chance to send your name on many deep space missions. My own name has been sent many times; most recently on the InSight mission that landed on Mars in 2018. When New Horizons flew past Pluto in 2015 this writer's name in digital form was on the spacecraft. When New Horizons flew past Ultima Thule in early 2019 a good luck message from me had been beamed across the Solar System. When the Cassini mission launched to Saturn in 1997 my name was on the spacecraft. It was also on the Huygens probe that floated down through the atmosphere of Titan in 2005 and now resides on the surface.

Readers wanting to have their name attached to planetary missions should visit the mission websites as launch approaches. The next NASA mission to Mars is scheduled for launch in the summer of 2020.

When the time comes for me to leave this mortal life, I doubt if I shall get my ashes blasted in the general direction of Alpha Centauri, though my name is on a spacecraft that is leaving the Solar System. Perhaps instead I could have my ashes scattered at Cape Canaveral Air Force Station in Florida where many planetary missions started their journey. Now there's a thought. Must get my will changed!

March

Full Moon: 9 March
New Moon: 24 March

MERCURY is barely visible in the morning sky this month in northern temperate latitudes. However, the elusive planet rises nearly 27° above the dawn horizon for those observing from the southern hemisphere. Mercury resumes direct motion on 9 March and reaches greatest elongation west (the largest of the year at 28°) on 24 March. Three days later Mercury is at aphelion. The closest planet to the Sun starts the month at magnitude +3.6 but ends at a much brighter +0.1.

Morning Apparition of Mercury
25 February to 4 May

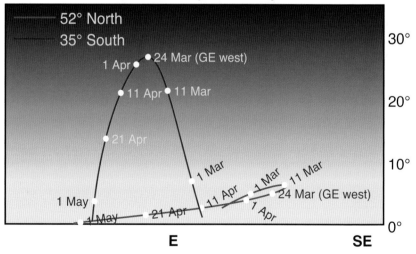

VENUS had a close encounter with Neptune in January and now it is the turn of Uranus when on 8 March, the two planets are only 2.2° apart. Because the evening sky may not be fully dark, optical aids will be required to see sixth-magnitude Uranus. Venus reaches perihelion on 20 March and four days later, attains greatest elongation east of $46°$. Over the course of the month, Venus changes from a waning gibbous object to a waning crescent and brightens slightly from -4.2 to -4.4 at the same time. The evening star is best viewed from northern latitudes where it reaches around $40°$ or more in altitude above the western horizon. Southern hemisphere observers have a much poorer apparition, with Venus slowly declining above the north western horizon.

EARTH reaches an equinox on 20 March, marking the beginning of astronomical spring in the northern hemisphere and astronomical autumn in the south. Four days later the Moon is at its farthest apogee of the year.

MARS is again occulted by the waning crescent Moon, this time on 18 March. However, the actual occultation event, which starts around 06:30 UT, is visible only from the southern tip of South America and the Antarctic. Two days later, the red planet is found $0.7°$ south of the considerably brighter Jupiter (magnitude $+0.9$ versus magnitude -2.1). A similar close encounter with Saturn occurs on 31 March, the day after Mars crosses from Sagittarius to Capricornus. This congregation of the three bright superior planets is best seen from southern latitudes where the trio rise around midnight or a little before. Observers in northern temperate latitudes will need to make an early start, with Mars rising just before dawn.

JUPITER is best viewed from the southern hemisphere where it begins to rise before midnight. Unfortunately for observers in northern temperate latitudes, twilight begins soon after the planet rises in the early morning hours. Shining brightly at magnitude -2.0 in Sagittarius, it is found $1.5°$ north of the waning crescent Moon on 18 March. Two days later, Jupiter and first-magnitude Mars are only $0.7°$ apart.

SATURN was repeatedly occulted by the Moon last year and continues to have close encounters with our satellite in 2020. On 19 March, the ringed planet is $2.1°$ north of the waning crescent Moon in Sagittarius. It comes even closer to Mars on the last day of the month when the planets are less than a degree apart. Between these two events, on 21 March, Saturn crosses the border into the constellation Capricornus. Saturn rises just before the onset of morning twilight for observers in the northern hemisphere but viewers in southern latitudes can see the planet from around midnight onwards.

URANUS is in the company of Venus on 8 March when the green ice giant and the evening star close to within 2.2° of each other. Planet watchers in northern temperate latitudes have much the best view of this faint object as the ecliptic is reasonably high above the horizon during the evening. A nearly Full Moon will brighten the sky significantly so optical aids will be necessary to see this sixth-magnitude planet.

NEPTUNE reaches conjunction on 8 March and is unobservable this month.

Pre-Iron Age Uses of Meteorites

Carolyn Kennett

A small hut discovered by archaeologists in 2014 near Lake Świdwie in north western Poland contained a variety of talismans dating from around 9,000 years ago. Three iron stones were found among the fascinating remains. Small and cylindrical in shape the stones were very heavy for their size. These stones from Poland are not the only example of iron objects being shaped by humans before

The Alaca Höyük dagger, with iron blade and golden hilt, dating to around 2,500 bc. (Wikimedia Commons / Stipich Béla)

the Iron Age. This is curious as iron metallurgy involves complex processes and the smelting of iron didn't become widespread until the Iron Age or approximately 1,200 BC. There are increasing examples of finds of iron objects which predate this period, but they are rare and come from a wide number of worldwide locations. The main explanation for the source of iron for these objects is that it originated from meteorites.

A meteorite fall – accompanied with displays of light and sonic sounds – would have made a deep impression on any who witnessed it. Recovery of an accompanying stone or iron meteorite would have been revered perhaps as a gift from a higher god. But it wouldn't have been necessary for people to witness the fall to think that the meteorite was special. Most meteorites are stony so a find of an even rarer heavy iron object on the ground would have been treated as significant. There are many examples of societies who regarded them as a valuable source of rare metal. Examples of objects which predate the Iron Age but are made from iron include a set of heavy balls from Tepe Sialk, Iran (4,600–4,100 BC); beads from Gerzeh, Egypt (3,200 BC); a dagger from Alaca Höyük, Turkey (2,500 BC); an axe from Ugarit, northern Syria (1,400 BC);

Albert Jambon evaluating meteorite content with the X-Ray Spectrometer. (B. Devouard)

two axes from the Shang dynasty civilization, China (1,400 BC); and the dagger, bracelet and headrest of Tutankhamun (Egypt, 1,350 BC).

In 2017, the French scientist and mineralogist Albert Jambon at the Institut de Minéralogie, de Physique des Matériaux et de Cosmochimie, evaluated a number of these iron objects using an X-Ray fluorescence spectrometer. The examination included the beads from Gerzeh and the Tutankhamun dagger, amongst others. His investigation of the iron/nickel content of these objects was compared to a large sample of meteorites. The analysis found that the high nickel content was consistent with the objects being from a meteorite source. He concluded that none of the iron objects tested was from a terrestrial source of iron.

There are examples of societies which utilised large meteorites as a collective. In the Cape York region of Greenland, a meteorite known locally as the "Ahnighito" (one of the largest iron meteorites in the world) was harvested by the Inuit. They used stone tools to chip away at this large metal object, allowing them to access metal tips for harpoons before iron was brought from neighbouring Iceland. The fall of this meteorite is dated to 10,000 BC, well before human occupation of the country, but its find and subsequent use demonstrates the ingenuity of the local population.

The three iron stones from Poland are yet to be confirmed as meteoric in origin. They predate any other finds by millennia and, if discovered to be from a meteorite, they show how long humanity has had a relationship with these wonderful objects.

This is just an introduction to the topic of people's relationship with meteorites; a more detailed examination will appear in a forthcoming edition of the Yearbook of Astronomy.

April

MERCURY is present in the morning sky, well-placed in the east for planet watchers in equatorial and southern latitudes but very near the horizon for those situated farther north. The zero-magnitude object is only 1.3° south of eighth-magnitude Neptune on the fourth day of the month. Mercury brightens steadily this month, from magnitude +0.1 to −1.6 by the end of April, but also gets nearer to the horizon as it approaches superior conjunction in May.

VENUS is 0.3° south of the open star cluster, the Pleiades (M45), on the third day of the month. The evening star is getting closer to Earth and growing larger in telescopes; at the same time, the illuminated portion of the planet is getting smaller. However, the planet continues to brighten, beginning the month at magnitude −4.4 and ending at −4.5. Venus is losing altitude but is still particularly well-placed for viewing from the northern hemisphere.

EARTH enjoys the light of a Super Moon on 8 April. Because the Moon's orbit around the Earth is elliptical, the apparent angular diameter of our satellite is constantly changing. The closest perigee of this year occurs just nine hours before Full Moon on the eighth. This means that of all the Full Moons of 2020, April's is the nearest one to the Earth and thus, has the largest apparent angular diameter. This phenomenon is sometimes referred to as a Super Moon.

MARS is in Capricornus throughout April, brightening from magnitude +0.8 to +0.4 over the course of the month. The waning crescent Moon passes 2.0° south of the red planet on 16 April. The ecliptic is low to the southern horizon for those in northern temperate latitudes so Mars never gains much altitude; it rises with the dawn. However, Mars rises before midnight for observers in the southern hemisphere and is high overhead when morning twilight begins.

JUPITER appears 2.0° north of the waning gibbous Moon on 14 April and on the next day attains west quadrature, when Jupiter–Earth–Sun makes a right angle. The gas giant rises late in the evening for favoured southern observers but remains confined to morning skies for those in northern latitudes. Jupiter's magnitude increases from −2.1 to −2.3 this month. Look for the bright planet in Sagittarius.

SATURN is in the zodiacal constellation of Capricornus this month. Although the rings are continuing to close, the planet is getting closer to Earth so its brightness is fairly constant at +0.7. On 15 April, the Moon passes 2.5° south of Saturn. West quadrature is on 21 April. Southern hemisphere observers have much the best views, with Saturn rising late in the evening. The ringed planet remains strictly a morning sky object for those looking for it from northern latitudes.

URANUS is at conjunction on 26 April and is lost to view in the Sun's glare.

NEPTUNE is now past conjunction and has entered the morning sky. Observers in southern latitudes will have the best opportunity to see Neptune just 1.3° north of Mercury in the east before sunrise. A small telescope offers the best chance to spot eighth-magnitude Neptune in the dawn sky but as always, exercise extreme caution when observing any celestial object in the vicinity of the Sun.

The Revd Doctor William Pearson, Co-Founder of the Royal Astronomical Society

Mike Frost

William Pearson was born on 23 April 1767 at Whitbeck, Cumberland in the southern Lake District, from a family of yeoman farmers. The accomplished telescope maker Edward Troughton, 12 years Pearson's senior, came from a neighbouring village, and we can speculate that he may have sparked an interest in astronomy in William. Pearson attended Hawkshead Grammar School, where a fellow pupil was the poet William Wordsworth.

After a brief stint as an assistant schoolmaster at Hawkshead, the next place we find Pearson was Lincoln, where he was a schoolmaster, and also ordained to the ministry. Around this time he married Frances Low. They had a daughter, also Frances, and you can see a portrait of the family on the wall of the Fellows Room in Burlington House, Piccadilly, home of the Royal Astronomical Society, in which Pearson is demonstrating to his family one of the models he had designed (an orrery) to demonstrate the workings of the solar

Portrait of Revd Doctor William Pearson, from the *History of the Royal Astronomical Society (1820–1920).* (Wikimedia Commons)

system. In Lincoln he was known for his public lectures on astronomy; also for the articles he wrote for Rees's Cyclopedia on a variety of topics concerning astronomical and horological instruments.

The next move in his career was to London, as partner in a boys' preparatory school in Parson's Green, Fulham; in parallel he became rector of Perivale, Fulham. He then founded his own school, Temple Grove, in East Sheen, Surrey. East Sheen was the location of his first observatory, featuring a revolutionary rotatable dome built by John Smeaton, of lighthouse fame. The school was very successful and made Pearson's fortune; his pupils included the Duke of Wellington's son. Pearson became a respected member of London society.

In the opening decades of the nineteenth century, the idea had been growing for a London astronomical society. Pearson had suggested it in letters, as had Francis Baily. In 1819 the two got together and on 12 January 1820, the Astronomical Society of London was founded in the Freemasons' Tavern, with 14 founder members, including John Herschel and Charles Babbage, and Pearson as treasurer. King George IV was invited to be patron, and the society eventually received its charter and became the Royal Astronomical Society in 1831.

A front view of Pearson's observatory in South Kilworth, now a private house. (Mike Frost)

Pearson was at the height of his success, with a profitable business and connections to the heart of London society. Yet in 1821 he sold his interests in the school, and took up the rectorship of South Kilworth, a small village in rural Leicestershire, which he had held in absentia since 1817. Why did he do this? The answer was surely that Pearson wanted to concentrate on his observations. Pearson built a new wing for the rectory, to house transit telescopes, with a more moveable refractor in a summer house in the garden. This arrangement proved unsatisfactory because of smoke from other houses in the village, so in 1834 Pearson built himself a new observatory on the edge of the village, with a clear south horizon.

From this observatory he carried out decades of observations of stellar positions, producing a catalogue of 520 stars which could be occulted by the Moon; and a long series of mid-day observations of the Sun, from which he derived a more accurate value of the obliquity of the ecliptic, the angle of tilt of

The boathouse which Pearson built on the shores of Grasmere, and which Wordsworth did not like! The plaque says "W.P. 1843". (Mike Frost)

the Earth's axis. This was cutting-edge astronomy for the time, using first-rate equipment. One of his transit scopes, for example, was originally destined for the Imperial Observatory, St Petersburg, before Napoleon's invasion of Russia made export impossible. St Petersburg's loss became South Kilworth's gain!

Pearson was generous about passing on his observing techniques, producing a two volume work, *Practical Astronomy*, which won the RAS's gold medal, and was still regarded as a relevant textbook at the turn of the twentieth century.

Both the rectory and observatory are still extant, though both are now private dwellings. The observatory had a chequered career, being used as a store, and then a cowshed, before being restored in recent years. A slate sundial (for ornamental not practical use) from the observatory is now on display in Market Harborough museum. A few of Pearson's instruments still exist – for example the author has seen a satellitium (clockwork model of Jupiter and its satellites), built to Pearson's design, which is held in storage at the Museum of the History of Science in Oxford. A rather larger construction was a slate boathouse, built by Pearson on the shores of Grasmere in the centre of the Lake District, where he owned land. This attracted the ire of Wordsworth, who called it "a tasteless thing in itself". In truth, the boathouse is under-stated and in harmony with its surroundings.

But Pearson's lasting legacy is the society he helped found, and remained an active member of until his death in 1847. The Royal Astronomical Society is 200 years old in 2020, and we would do well to remember the schoolmaster and clergyman who played such a fundamental role in its foundation.

References

Revd. Dr William Pearson (1767–1847): a Founder of the Royal Astronomical Society, S.J.Gurman and S.R.Harratt (Q.J.R. Astr. Soc. (1994) 35, 271–292)

Reverend Doctor William Pearson in South Kilworth, Leicestershire, M. A. Frost (The Antiquarian Astronomer (2006) 3, 49–56)

May

Full Moon: 7 May
New Moon: 22 May

MERCURY comes to within 0.3° of Uranus on the opening day of May but with superior conjunction only three days later, the event is unobservable. The tiny planet quickly reappears in the west after sunset in what is the best evening apparition of the year for northern hemisphere observers. Perihelion occurs on 10 May. Mercury pairs up with Venus on 22 May when the two planets are only 0.9° apart, 19° from the Sun. Mercury is a bright −0.6 magnitude but Venus easily dominates at −4.2. Mercury dims to magnitude +0.2 by the end of the month.

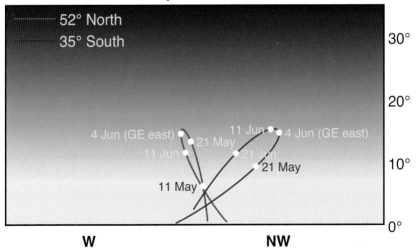

Evening Apparition of Mercury
4 May to 30 June

VENUS is approaching inferior conjunction and that means it must enter retrograde motion before that event. In 2020, Venus reaches this stationary point on 13 May. Mercury and Venus come to within 0.9° of each other on

22 May but the evening star is getting low to the horizon by this point. It all but disappears in the evening twilight by the end of the month.

MARS leaves Capricornus for Aquarius on 8 May. One week later, it is 2.8° north of the waning crescent Moon. Best seen from southern latitudes where it rises before midnight, Mars remains a morning sky object for northern viewers and stays close to the horizon. The red planet is at magnitude zero by the end of the month.

JUPITER has a close encounter with the waning gibbous Moon on 12 May, appearing 2.3° north of our satellite. Two days later it reaches a stationary point and begins retrograde motion. Jupiter continues to brighten from magnitude −2.3 to −2.6 over the course of the month. It is best seen from the southern hemisphere where it appears above the horizon in mid-evening. Planet watchers farther north must wait until around midnight before Jupiter rises in Sagittarius.

SATURN reaches declination −19.9° on 6 May, the farthest north it gets this year. Five days later, it reverses course and heads back toward Sagittarius although it won't re-cross the border into that constellation until July. Saturn is slowly brightening as it approaches opposition and shines at magnitude +0.6. This is despite the rings closing to their minimum tilt of the year (20.5°) on 8 May. Southern latitudes are favoured for spotting this planet which rises mid-evening. However, observers in northern temperate latitudes will not be able to view the ringed planet until around midnight.

URANUS was at conjunction late last month, so its close encounter with Mercury on the first day of May is unobservable with the two planets only 4° from the Sun. Now a morning sky object, Uranus slowly distances itself from our star and emerges from the dawn twilight by the end of the month.

NEPTUNE shines at magnitude +7.9 amongst the dim stars of Aquarius. The faint planet is difficult to observe from the northern hemisphere as Neptune rises during morning twilight but it is well aloft in dark skies as viewed from southern latitudes.

Astronomy Stamps

Martin Beech

Collecting – it is a human compulsion. We have evolved, for so it would seem, to gather and measure, record and describe. Astronomers are no less caught-up in this obsession to generate lists and tick off objects observed – for what else is a star chart? Likewise, how many times, within the gathered hush of an astronomy club, has one heard in exalted tone, "Messier marathon tonight"? We collect and we catalogue. The ancients were but little different to us, although they did have the good fortune to be born at a time when the stars needed naming. The brighter stars and the many star-groupings that caught their eyes, and their imaginations, are now our astronomical heritage, and we still follow the annual and nightly progression of the pointillist hunters, mythical beasts, arthropods, whales and Polaris-circling bears that inhabit the sky.

Indeed, there is great delight in having once seen an object or recognized some pattern, to name and then catalogue it, and to eventually contemplate its *genus loci*. For the active astronomer a clear sky is essential in order to see

 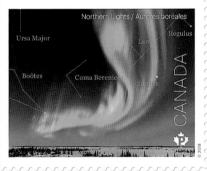

Astronomy stamps issued by the Canadian Post Office in 2018. At left the stamp shows the sky constellations towards the galactic centre. At right an auroral curtain is shown against the backdrop of the constellation of Leo. (Image © Canada Post 2018)

and record the sky, and this precondition, as we all know too well, is often denied us. But, as the weather can turn against our astronomical forays, so our thoughts can turn inwards and contemplate the indoor pursuits of collection and discernment. Indeed, the indoor astronomer has many choices between colour-bursting texts and magazines, the internet, computer apps, and the television to engage their attention; but best of all are astronomy stamps. Stamps, like the stars and constellations, are there to be selected and sorted, catalogued and contemplated.

The act of collecting stamps provides for a visual history of the sciences and a rogue's gallery of those famous scientists that have made extraordinary contributions to human knowledge. They evoke the *narrativium* latent within the history of science, and they remind us of the greatness of human ingenuity. Stamps convey a message that has an historical, cultural, philosophical, and political context. Indeed, stamps are a form of semiotic communication, working both as a mirror to reflect human ideals and as a lens to celebrate discovery and discoverers – and, if one wants, they have a practical value too. Stamps are visual time capsules, small snippets of celebration, with both national and international parties commemorating past moments of human greatness. In the words offered by physicist Hans Bethe, stamps are inherently peaceful and positive, and very few stamps have even been produced that portray or attempt to celebrate natural disasters and/or the less than noble side of human innovation. Stamps are additionally interesting as tactile objects that can portray numerous polygonal forms, various textures and highly inventive artwork. Stamps have been produced from paper, cork, plastic and tinfoil; they have contained holograms, they have been printed with heat sensitive colour-changing ink, and with ink infused with meteorite dust; they have been produced with elaborate embossing and even with scratch and sniff fragrances – some have additionally shown

The first astronomy-themed stamp was issued by the Brazilian Post Office in 1887 and featured the iconic southern constellation of the Crux. (PD-BrazilGov)

Finding the South Celestial Pole with the Southern Cross (Crux) and the pointers α and β Centauri. From the A to Z of New Zealand series of stamps issued by the New Zealand Post in 2008. (Image © New Zealand Post 2018)

The Ring Nebula (M57) in Lyra as depicted in the six-stamp series issued by the Mexican Postal Service in 1942 celebrating the opening of the Tonantzintla Observatory. (PD-MexicoGov)

unintentional printing errors which transforms them from a common place object into something that is both unique and highly collectable. It has been estimated by Australian writer Christopher Yardley that over 600,000 stamp designs have been issued since 1840, and that about 10 percent of these issues relate to the sciences and the celebration of famous scientists.

While the first Penny Black stamps were issued in Great Britain in May of 1840, the first stamp to appear with an astronomy motif was that produced by the Brazilian Post Office in 1887. This stamp featured the iconic southern constellation of Crux – a constellation that has gone on to be featured on many southern hemisphere stamps ever since. The first science-themed stamps began to appear in the 1920s and '30s, with the first stamp of an astronomer, appropriately Copernicus, being printed by the Polish Postal Office in 1923. The first astrophysics-related stamps, a set of several deep-sky images (and a Hertzsprung-Russell diagram) celebrating the opening of the Tonantzintla Observatory, were printed by the Mexican Postal Service in 1942. The first science-related stamp to be printed in England appeared in 1965 and it celebrated the pioneering development of antiseptic surgery by Joseph Lister. However, in the following year the first astronomy-related stamp was issued in a series of stamps themed according to the alphabet. Indeed, under the entry for J was shown the Jodrell Bank (now the Lovell) radio telescope. British astronomy was further celebrated in 1970 with a stamp commemorating the 150th anniversary of the Royal Astronomical Society – this stamp appropriately showing William Herschel, Francis Baily and John Herschel set against a view of Herschel's great 40-foot reflector. The Royal Greenwich Observatory and the 100th anniversary of the international establishment of the Prime Meridian were further celebrated by the Post Office in 1984. Not missing the chance to celebrate the great Edmund Halley, a set of stamps (with art work by illustrator Ralph Steadman) was issued to coincide with the return of his famous comet in 1986. The long history and heritage of British astronomy was further highlighted in a four-stamp series published by the Post Office in 1990 – this set showing stylized images of observatories, telescopes, navigation instruments, and the astronomical sky. The 50th anniversary of Sir Patrick Moore's *The Sky at Night* television programme was additionally celebrated by a series of stamps issued in 2007.

The great New Zealand born physicist Ernest Rutherford is generally given credit for the quip that "all science is either physics or stamp collecting". Usually interpreted as a disparaging slight to both the art of stamp collecting and the non-physics-based sciences, we prefer to remember the comments attributed to Danish quantum mechanics pioneer Niels Bohr, who reasoned that a new scientific result is like, "an avid stamp collector who had just gotten his hands on a rare item for the collection". Indeed, stamps and science are all about discovery and celebration. Beyond the written word and mathematical formulae, however, stamps provide a visual, political, social and historical account of human achievement. They mark special moments in science and they allow us to celebrate the lives of those famous scientists (and dreamers) that have literally changed our worldview. There may be no calculus of stamps or functional analysis of stamp collecting, but stamps are assuredly a vital part of the *story* of astronomy and of human scientific exploration.

Further reading

Astronomy and Cosmology in Stamps: **http://ircamera.as.arizona.edu/NatSci102/ NatSci/images/extstamps.htm**

Martin Beech: *Cigarette and trade-card astronomy: c1900–c.2000* – a journey from engaged imagination to passive data collection. The Observatory (137, 288, 2017)

Renato Dicati: *Stamping Through Astronomy.* Springer, Milan (2013)

Ian Ridpath: **www.ianridpath.com/stamps/stampindex.htm**

Morning Apparition of Venus
June 2020 to March 2021

June

Full Moon: 5 June
New Moon: 21 June

MERCURY is at greatest elongation east (24°) on 4 June, reaching its maximum altitude in the west at around the same time. It spends the rest of June descending back toward the horizon, disappearing in the evening twilight by the end of the month. The planet reaches a stationary point on 17 June and begins retrograde motion. Six days later, Mercury is at aphelion. Inferior conjunction occurs next month and so the planet dims rapidly, ending up as a sixth-magnitude object by the end of June.

VENUS reaches inferior conjunction on 3 June. It reappears at dawn, assuming the mantle of the morning star for the rest of the year. As with the evening apparition that has just ended, this morning apparition favours the northern hemisphere, with bright Venus attaining nearly 50° in altitude later in the year as seen from northern equatorial latitudes. The waxing crescent Venus is 0.7° south of the waning crescent Moon on 19 June and very early risers in north eastern North America may witness a lunar occultation around 07:00 UT. Venus resumes direct motion on 24 June.

EARTH sees the second of four lunar eclipses this year when the Full Moon dips into the Earth's penumbra on 5 June. The first solstice of the year occurs on 20 June when summer arrives in the northern hemisphere and winter takes hold in the south. The following day brings an annular solar eclipse to parts of Africa, the Middle East and southern Asia. Both eclipses are discussed in the section *Eclipses in 2020*.

MARS is best observed in the morning hours although for astronomers in the southern hemisphere, the red planet rises in late evening. It reaches west quadrature on 6 June and one week later, has close encounters both with the Last Quarter Moon and Neptune. On 24 June, Mars crosses from Aquarius into the constellation of Pisces. With opposition approaching later this year, Mars draws nearer to Earth and brightens from magnitude 0.0 to −0.5 by the end of June.

JUPITER encounters the waning gibbous Moon on 8 June, appearing 2.2° north of our satellite. Now blazing away at magnitude −2.6 in Sagittarius, the gas giant rises in early evening for planet watchers in the southern hemisphere. Northern observers are finally able to view Jupiter low in the east late in the evening.

SATURN is in retrograde in Capricornus, brightening from magnitude +0.5 to +0.3 at the end of the month. The waning gibbous Moon passes less than 3° south of the planet on 9 June. It is best seen from southern latitudes where it rises in the early evening, with northern observers only now able to see the planet appear in the east before midnight.

URANUS rises in the early morning hours, appearing low in the east as seen from the northern hemisphere and in the northeast when viewed from southern latitudes. Look for it, shining dimly at magnitude +5.9, in the faint constellation of Aries.

NEPTUNE reaches west quadrature on 11 June, at an elongation of 90° from the Sun. Mars moves past two days later, the two planets only 1.6° apart at their closest approach. However, although Mars is a bright zero-magnitude object, Neptune is at best magnitude +7.9, necessitating optical aids to see it. The blue giant reaches its maximum declination south for the year (−4.7°) on 19 June and begins retrograde motion four days later. Neptune rises around midnight in Aquarius.

Cometary Comedy and Chaos

Courtney Seligman

Comets have long been considered as omens and, the gods being known to be fickle, usually evil omens. For the English King Harold the 1066 apparition of what we now call Halley's Comet certainly proved an evil omen, as shown in the Bayeux Tapestry, which was produced for the victorious Normans. This article concerns other cometary events that appeared to bode very ill.

In 1848 the English astronomer John Russell Hind revisited a long-held theory that the Great Comet of 1556 (often known as the comet of Charles V since it appeared during the reign of Holy Roman Emperor Charles V) was a reappearance of the Great Comet of 1264, and concluded that its return could be expected between 1848 and 1860. In an 1855 review of comets of the 19th

The Bayeux Tapestry depicts the appearance of Halley's Comet, and terrified people regarding it as an evil omen. (Wikimedia Commons/Myrabella)

century, the French astronomer and physicist Jacques Babinet discussed Hind's calculations and wrote that, since the comet had not yet appeared, it was likely to reappear between 1856 and 1860 (give or take 10 years). Due to apocalyptic articles printed in 1855 and 1857, it soon became "common knowledge" that a "German astronomer" claimed that the comet would reappear in June of 1857. Rumours and gossip subsequently turned that into a prediction that the comet would strike the Earth on 13 June 1857, ending all life on our planet.

John Russell Hind, whose erroneous research led to the creation of "The Comet That Never Was". (Wikimedia Commons / Henry Joseph Whitlock)

The resulting fear spanned Europe and the Americas, although for some reason the greatest terror was felt by those living in or near Paris. The French artist and caricaturist Honoré Daumier took advantage of this to publish a series of satiric cartoons in *La Charivari* (a French publication similar in tone and content to the English *Punch* and published in Paris from 1832 to 1937), poking fun at the Parisians, the "German astronomer" and Monsieur Babinet (perhaps because he was a well-known lecturer on astronomy). As it turned out, the comets of 1264 and 1556 were not the same object, our continued existence proves there was no comet that ran into the Earth in 1857, and Hind's comet is now referred to as "The Comet That Never Was".

However, the very next year saw the appearance of Donati's Comet, the second brightest comet of the 19th century. Seizing a last opportunity to poke fun at the unfortunate Jacques Babinet, Daumier published a cartoon of him still looking for the nonexistent 1857 comet while his housekeeper frantically points out Donati's Comet. (Although none of Daumier's cartoons are shown here, they may be seen at the Daumier Register or on my website at **cseligman. com**). Donati's Comet was not only exceptionally bright, but passed well to the north of the ecliptic and became easily visible in the northern sky, especially during September and October of 1858. Many observers (including Abraham Lincoln, who observed it the night before his third debate with Stephen Douglas)

Donati's Comet recorded near the bright star Arcturus in the constellation Boötes on 15 October 1858. (Wikimedia Commons/Edmund Weiß – *Bilderatlas der Sternenwelt*)

considered it the most beautiful comet they had ever seen and pictures of it were still in wide circulation in the early 20th century.

Although catastrophic encounters with comets are rare, the Earth does run into small bits of comets (namely, particles from their dust tails) every day. The Earth's orbital motion sweeps up about 5 to 10 tons of meteoric material daily, 90% of which is comprised of cometary remnants, and the meteors that result from such encounters are therefore almost always "cometary" collisions with our atmosphere. A few comets have "dust" trails that pass close to the orbit of the Earth so that each year, when the Earth is nearest the point in its orbit where such a dust trail nearly intercepts our orbit, meteor showers occur. Halley's Comet is responsible for two such showers – the Eta Aquarids and Orionids. This is due to the fact that the orbit of the comet passes near to that of our planet, both on its fall towards perihelion and on its subsequent path away from the Sun.

Another comet which produces an annual meteor shower is Comet Tempel-Tuttle, which gives rise to the Leonids. This is a shower that on some occasions can produce spectacular meteor storms, one of which caused a panic similar to that of 1857.

On the night of 18 November 1833, the eastern half of the United States, and in particular the south eastern part of the country, was witness to the greatest meteor storm in recorded history. At its peak, up to 100,000 meteors per hour were seen streaking across the sky, and hundreds of thousands of people (including the 24-year-old Abraham Lincoln) gazed in awe at the sight, while tens of thousands of others were terrified by the phenomenon, thinking that the Day of Judgment was at hand, and praying for all they were worth. For the best part of a century many southerners referred to 1833 as "the year that stars fell", and in 1934 a book of anecdotes collected from Alabamians was titled *Stars Fell On Alabama* because of the tales of soul-searing terror experienced by the ancestors of those reciting the anecdotes. There is little doubt that the 1934 hit song of the same name was based on the title of the book, and though the song does not refer directly to the meteor storm it carries the memory of that storm into the future.

A similar but weaker Leonid storm took place in 1866, and it was soon realized that the period between the 1833 and 1866 meteor storms was the same as the orbital period of Comet Tempel-Tuttle, which had been discovered near the beginning of 1866. The orbit of this comet was found to almost exactly intersect the Earth's orbit, and another such storm was anticipated in 1899, although in the interim the orbit of the comet was changed by a relatively close encounter with Jupiter. As a result, no such storm occurred in either 1899 or 1933. However, in 1966 there was an unexpected repeat (though on a much smaller scale) of the storms of the 1800s, and in the decades since then it has been realized that even though the comet's orbit had been changed, dust trails left by its passages hundreds or even thousands of years ago can still produce unusually intense meteor showers at intervals of about 33 years.

Fortunately, such meteor showers and storms are about the only real encounters we have with comets on a short-term basis. The Earth collides with objects a few metres across almost every week (the United States has shared software with other countries that allows them to tell that the resulting meteoric explosions were not caused by atomic weapons), and with objects up to a few tens of metres across every few decades or centuries (such as the

The Leonid Meteor Shower of 1833 (based on an eyewitness account) as depicted on an engraving from *Music of the Spheres* by Florence Armstrong Grondal. (Wikimedia Commons)

Tunguska fireball of 1908 and the Chelyabinsk meteor of 2013). But impacts by objects hundreds of metres across, such as the iron meteoroid that produced Meteor Crater in northern Arizona, only occur about every 50 to 100 thousand or so years, and objects big enough to cause truly catastrophic events, such as the roughly 10 kilometre wide asteroid that struck what is now Yucatan about 65 million years ago (and ended the age of dinosaurs) represent a once in two hundred million years' event. So catastrophic events can occur, but only at such long intervals that the chances of one occurring during the life of anyone now living are very close to zero. And, although social media allow the spreading of false rumours of astronomical catastrophe on a continual basis, one presumes that the readers of this volume have more sense than to believe in such nonsense.

July

Full Moon: 5 July
New Moon: 20 July

MERCURY is at inferior conjunction on 1 July, passing into the morning sky. It resumes direct motion on 12 July and ten days later, reaches greatest elongation west of 20°. The planet's brightness varies greatly this month, beginning at sixth magnitude and ending July at −0.7. This dawn apparition in the east favours equatorial regions.

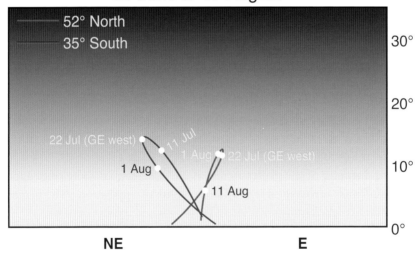

Morning Apparition of Mercury
30 June to 17 August

52° North
35° South

30°

20°

22 Jul (GE west)   11 Jul   1 Aug   22 Jul (GE west)   10°

1 Aug

11 Aug

0°

NE   E

VENUS continues its morning ascent above the eastern horizon and is best seen from northern latitudes. It shines at magnitude −4.5 and is easily the brightest non-lunar object in the dawn sky. It reaches aphelion on 10 July and is found 1.0° north of Aldebaran, the first-magnitude star of Taurus, the following day. The waning crescent Moon passes 3.1° north of Venus on 17 July. When viewed through a telescope, Venus appears as a waxing crescent, filling out from 19% to 43% over the course of the month.

EARTH is at its most distant from the Sun on 4 July when it reaches aphelion (1.017 AU or 152,100,000 km). The following day brings yet another penumbral lunar eclipse of the Full Moon, the details of which are outlined in the section *Eclipses in 2020*.

MARS spends the rest of the year in Pisces, apart from a short excursion (8–16 July) into the non-zodiacal constellation of Cetus. The waning gibbous Moon passes 2.0° south of the planet on 11 July. The red planet continues to brighten, going from magnitude −0.5 to −1.1 by the end of the month. It rises in mid- to late-evening and is most easily viewed from southern latitudes where it climbs high into the sky.

JUPITER is less than 2° north of the Full Moon as our satellite undergoes a penumbral lunar eclipse on 5 July but the main event occurs on 14 July when the gas giant arrives at opposition in the constellation of Sagittarius. On this day, Jupiter is 4.1 AU distant, appearing nearly 48 arc-seconds in diameter and shining at magnitude −2.8. Because it is opposite to the Sun in the sky this month, Jupiter rises at sunset and sets at sunrise.

SATURN is in retrograde motion and after 3½ months in Capricornus, returns to Sagittarius on 3 July. Three days later it has a close encounter with a nearly Full Moon but enjoys dark skies when it comes to opposition on the same day as New Moon on 20 July. At this time, Saturn is at a distance of 9.0 AU and shines at magnitude +0.2. The rings are tilted at an angle of 21.7° and the apparent diameter of the planetary disk is 18.5 arc-seconds. (The rings appear 42.1 arc-seconds across, nearly as wide as Jupiter at opposition.) The planet is on display all night, rising at sunset and setting when the sun appears.

URANUS passes into the evening sky by the end of the month but is best seen after midnight after it gains some altitude. At magnitude +5.8, Uranus requires dark skies and sharp eyes to spot it.

NEPTUNE has moved into the evening sky and now rises before midnight. Having entered retrograde motion late last month, it continues to backtrack across Aquarius. A small telescope is required to observe the magnitude +7.9 planet.

A Pair of Classic Doubles in Cygnus

John McCue

Observers at mid-northern latitudes can see the constellation of Cygnus (the Swan) straddling the Milky Way high up in the southern sky during August evenings, its leading star Deneb lying a little to the east of the bright star Vega in Lyra. Deneb and the brilliant Vega (in Lyra) and Altair (in Aquila) together form the conspicuous and spectacular Summer Triangle.

The Milky Way passes through Cygnus and provides a marvellous backdrop to the double stars in the constellation, an irresistible incentive to find them in a telescope. Cygnus is also known as the Northern Cross, a fact not lost on the American astronomer Percival Lowell, who in 1844 wrote of the constellation that the countless splendours in the sky were '... *crowned by the blazing Cross hung high o'er all ...*' – praise indeed!

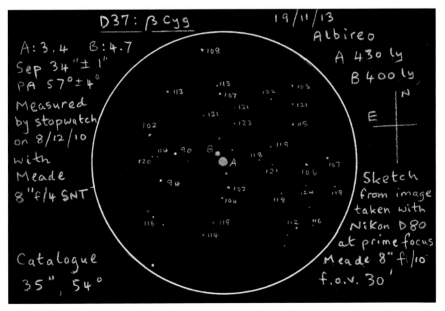

The author's observation of Albireo. Catalogue refers to the Washington Double Star Catalogue's current values of separation and position angle. The stopwatch method is described in detail in the *Yearbook of Astronomy 2019*. (John McCue)

Albireo (β Cygni)

The contrasting colours of this double star, located at the foot of the Northern Cross make it a magnificent spectacle. The primary star, at apparent magnitude 3.4, is yellow, and the secondary star, at 4.7, is blue. Try different eyepieces in your telescope to obtain the best view of these stellar colours. Burnham[1] recommends no more than a 30x magnification to see at its best the dazzling contrast of golden yellow and sapphire-blue. With your best field of view, The Milky Way background of scattered stars is apparent, augmenting this feast for the eyes.

These two stars are separated by 35 arc seconds and lie about 400 light years from us, but no-one is sure if they are actually orbiting each other. If they

An image of Albireo taken by Graham Darke of the Sunderland Astronomical Society, using a Canon 600 D camera attached to a 140mm triplet refractor at 980mm focal length. Each image is a stack of 20 exposures of 15 seconds duration. The short exposures help to preserve the colours in the stars. (Graham Darke)

do, it must take them many thousands of years as very little change in their configuration has been seen since the first measurement in 1755. Remarkably, the primary star has a companion, too close for the amateur telescope to detect, but it has been revealed by examining its spectrum, which is seen as composite. This yellow primary is of spectral type K3, and the huddled partner is type B9, the two spectra entwined as one. These two stars take about 100 years to orbit each other.

61 Cygni

At 11 light years, the two stars in this binary system were the first to have their distance measured. In 1792, the Italian astronomer Giuseppe Piazzi, first detected its unusually high proper motion around our Milky Way galaxy. Once news of this abnormally high motion got out, astronomers began referring to it as "Piazzi's Flying Star".

This image of 61 Cygni was obtained using the same equipment as that for Albireo (above). A drawing of 61 Cygni by the author appears in his article *Getting the Measure of Double Stars* in the *Yearbook of Astronomy 2019*. (Graham Darke)

The German astronomer and mathematician Friedrich Wilhelm Bessel reasoned that 61 Cygni must be situated fairly close to us to account for this speedy motion across the sky. In 1838, he took this a stage further when he succeeded in measuring the distance to this pair. It was the first time a stellar distance had been measured. Bessel used the parallax method – measuring the shift in position of the binary against the background of more distant stars as the Earth swings around to the other side of its orbit in a six-month period.

Although 61 Cygni is officially the first stellar distance measured, the historical background reveals that Thomas Henderson made observations of α Centauri from the Cape of Good Hope before Bessel, with the intention of ascertaining its distance. On his return to Scotland, as first Astronomer Royal for Scotland, he was inspired by Bessel's sensational success[2] to reduce his own observations and obtained α Centauri's distance. Around this time, another runner in the race, Friedrich Struve, a renowned double star observer, made a distance measurement of Vega, which unfortunately was not very accurate[3].

Bagnall[4] notes that these waltzing stars are visitors rushing through our neighbourhood from another part of our galaxy, and that in around 25 centuries from now they will be seen in the direction of Cassiopeia. The two stars are at apparent magnitudes 5.3[5] and 6.1[5] and are distinctly orangey. They orbit each other in about 660 years, the author measuring their separation at 31.7″ with error 0.2″ by a software analysis[6] of an image taken with the 2-metre Faulkes Telescope in Hawaii.

Notes

1. *Burnham's Celestial Handbook*, Volume 2, Robert Burnham, Jr., Dover Press, 1978.
2. *Flammarion Book of Astronomy*, Camille Flammarion, George Allen and Unwin Ltd., 1964.
3. *Data Book of Astronomy*, Patrick Moore and Robin Rees, CUP, 2011.
4. *The Star Atlas Companion*, Philip M. Bagnall, Springer, 2012.
5. *The Cambridge Double Star Catalogue*, James Mullaney and Wil Tirion, CUP, 2009.
6. *Yearbook of Astronomy 2019*, Brian Jones (Editor), White Owl.

August

Full Moon: 3 August
New Moon: 19 August

MERCURY rapidly loses altitude above the eastern horizon and vanishes by mid-month. After superior conjunction on 17 August, the closest planet to the Sun moves into the evening sky for what is the best apparition of the year for southern observers. However, this evening apparition is virtually unobservable for viewers in northern temperate latitudes, with Mercury never gaining more than a few degrees in altitude at sunset. Mercury is brightest at superior conjunction, when it isn't actually visible, but is a respectable magnitude −0.6 by the end of August. The planet also reaches perihelion early in the month, on the sixth.

VENUS is at greatest elongation west (46°) on 13 August. The morning star continues to gain altitude above the eastern horizon for observers in northern temperate latitudes but appears a little lower each day when seen from the southern hemisphere. Venus goes from the waxing crescent phase to the waxing gibbous phase this month but because it is drawing away from Earth, appears smaller in diameter. For this reason, it dims slightly, from magnitude −4.4 at the beginning of August to −4.2 by month's end.

MARS reaches perihelion on 3 August, the point in its orbit where it is nearest to the Sun. Six days later it is occulted by the waning gibbous Moon, an event visible from southern South America and Antarctica and beginning at approximately 07:00 UT. Its magnitude increases all month, ending at a brilliant −1.8 as opposition approaches. The red planet rises in mid-evening in Pisces and is best seen from southern latitudes.

JUPITER was at opposition last month so it is already above the horizon as the sky grows dark and does not set until after midnight. Found in Sagittarius shining at magnitude −2.7, it is best viewed from the southern hemisphere where the ecliptic is high in the sky. Jupiter has two encounters with the waxing gibbous Moon this month, on 1 August and again on 29 August.

SATURN is just past opposition and is a bright zero-magnitude object in Sagittarius. It is already aloft at sunset and sets as the dawn skies brighten. The ecliptic is low to the horizon as seen from the northern hemisphere so it never rises very high; the best observing opportunities of Saturn are from southern latitudes. The waxing gibbous Moon passes close by twice this month, on 2 August and again on 29 August.

URANUS reaches west quadrature on the second day of the month. This is when it is found 90° away from the Sun in the sky. On 14 August, the green ice giant reaches its maximum declination north for the year (+14.5°) and the following day, begins retrograde motion which will continue into next year. Rising in mid-evening by the end of August, Uranus is found in the zodiacal constellation of Aries.

NEPTUNE continues to rise earlier in the evening as it approaches opposition next month. Located in Aquarius, it is best seen from the southern hemisphere where it rises high into the sky after sunset.

Proper Motion

Martin Whipp

Nothing is Fixed

When describing retrograde planetary motion in his 1543 publication *De Revolutionibus Orbium Coelestium*, the Polish astronomer Nicolaus Copernicus was quoted as saying "All these phenomena proceed from the same cause, which is the Earth's motion. None of these phenomena appears in the fixed stars"[1]. For millennia it was regarded that the stars were just that – fixed. We now know differently of course; the Universe is expanding and everything is moving through space, sometimes in groups, sometimes individually. Within a human lifetime, the celestial sphere of stars above us appears relatively unchanged, but with accurate measurement we can see that some stars are indeed moving more than others relative to the Sun.

How It All Began

On 13 October 1892, the American astronomer and astrophotographer Edward Emerson Barnard became the first person to discover a comet by photographic methods (the comet in question was originally designated D/1892 T1, but is now known as 206P/Barnard–Boattini following its rediscovery by Italian astronomer Andrea Boattini in 2008). However, Barnard is best remembered for his work in measuring the proper motions of stars, which resulted in a ninth magnitude star in Ophiuchus bearing his name. Barnard's Star has the largest proper motion (10.3 arc-seconds per year) of any star known.

Before Barnard's observations, the star had been previously imaged, but never astrometrically measured. In 1916 Barnard calculated its proper motion and surmised that due to its high velocity and close proximity to the Sun that it was very probable that no other star would be found to have a larger proper motion. The Barnard's Star system is second in distance only to the Alpha Centauri system.

"S'cuse Me Mister …?"

As a public presenter of astronomy, there are a few popular questions that I get asked all the time: "What was that bright thing I saw last night?", "When will the Sun blow up?", "How do you go to the toilet in space?" You get the idea. But one common question can often be open to erroneous explanation, and it's this: "How long would it take to travel to Proxima Centauri in a spacecraft?"

At first glance this question seems to have a relatively simple answer. We know the distance to Proxima to be 4.24 light years, which is roughly 40,000,000,000,000 km. If we were to use the New Horizons spacecraft which sped through the Pluto system in 2015 at a velocity of 58,000 km per hour as our vehicle of choice, then simple division of these two figures tells us that it would take approximately 79,000 years to travel from Earth to Proxima Centauri. But that figure is actually incorrect. Why? Because of Proxima's proper motion.

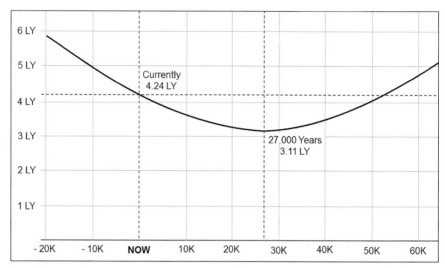

Graph showing the distance between the Sun and Proxima Centauri over a period of 80,000 years. Time is on the x-axis and distance in light years in on the y-axis. Note that the closest approach of Proxima occurs in 27,000 years from now, at a distance of 3.11 light years.

Astrometric measurements[2] of Proxima Centauri indicate that it is in an eccentric orbit around Alpha (α) Centauri. It has an orbital period of approximately 550,000 years. Currently it is moving towards the Earth at around 80,000km per hour. This means that in about 27,000 years time Proxima will only be 3.11 light years away. It will then start to recede once again.

So, regarding our imaginary interstellar journey to Proxima, the actual journey time is more like 65,000 years because our intended destination is courteously travelling towards us at the same time that we are travelling towards it. It is a bit like flying from London to Glasgow, but then discovering that Glasgow is in orbit around Edinburgh and by the time you get there, it's actually where you thought Newcastle was, and furthermore it once visited Durham!

Moving Next Door

There are many different ways that the stars have been catalogued over the centuries, but by far the most widely used (at least in amateur astronomy) is the Bayer system, named after the German astronomer Johann Bayer who devised the catalogue in 1603. Using this system, the stars in any particular constellation are given a Greek letter in descending order roughly according to their visual brightness. For example, Deneb is the brightest star in Cygnus, so its Bayer designation is given as Alpha (α) Cygni. Albireo is the second brightest, and so is known as Beta (β) Cygni and so on. However, Bayer did not strictly adhere to his rule of star brightness all of the time, sometimes preferring to use historical importance or the physical position within a constellation as a guide instead. As a result, there are over thirty constellations in which the star designated as Alpha is not the brightest one.

As if that wasn't complicated enough, some stars have altered their characteristics since the time of Bayer, which gives rise to further inconsistencies in classification. Perhaps the best known example is that for the majority of the time, within the constellation of Orion the brightest star is Rigel, but it is classified as Beta (β) Orionis. This is because Alpha Orionis (Betelgeuse) is a

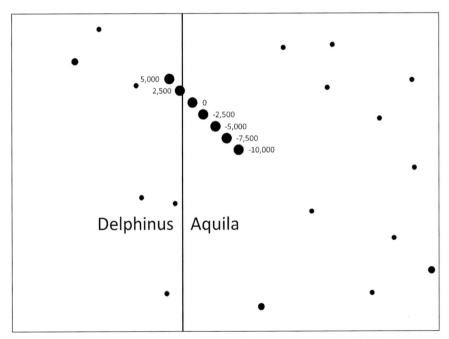

The proper motion of Rho Aquilae with positions shown at 2,500 year intervals. The star passed from Aquila into neighbouring Delphinus in 1992. Other surrounding stars had previously made the same move to the adjacent constellation, although with magnitudes of +8 and +9 none are bright enough to warrant a Bayer designation.

semi-regular variable star, and so occasionally spends a while visually brighter than Rigel.

But the strangest apparent misclassification is the story of Rho (ϱ) Aquilae, and this brings us back to proper motion. Rho Aquilae is a fifth magnitude A-class star which resides approximately 150 light years from Earth. Because of its relatively close proximity to us, combined with a receding velocity of 83,000km per hour[3], it has a proper motion of one arc second every 12.5 years[4]. Why is this so peculiar? Because in 1992, Rho Aquilae crossed the border from Aquila in to the neighbouring constellation Delphinus, becoming the first star ever to have a Bayer designation different to the constellation that it resides in.

Over the course of time this will occur more and more, and we can eventually expect a lot of interlopers from neighbouring constellations. Whether or not the International Astronomical Union will decide to rename them to reflect their new homes remains to be seen.

Notes

1. As quoted by Edwin Arthur Burtt in *The Metaphysical Foundations of Modern Physical Science* (1925) Book 1, Chapter 10.
2. *Proxima's Orbit around Alpha Centauri, Astronomy & Astrophysics.* Kervell, Thevenin, Lovis (2017)
3. *Catalogue of Fundamental Stars* (FK6) R. Wielen et al. (1999)
4. *Validation of the new Hipparcos reduction, Astronomy & Astrophysics* F van Leeuwen (2007)

September

Full Moon: 2 September
New Moon: 17 September

MERCURY is a zero-magnitude evening sky object, soaring high above the western horizon when viewed from southern latitudes. This apparition is a disappointment for observers in the northern hemisphere, with the planet remaining close to the horizon throughout. Mercury attains another aphelion on 19 September and three days later, is found 0.3° north of Spica, the alpha star of the constellation Virgo.

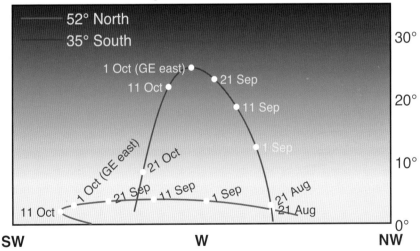

Evening Apparition of Mercury
17 August to 25 October

VENUS is the brilliant morning star. Now past greatest elongation, it slowly descends toward the eastern horizon and is best seen from the northern hemisphere. The phase increases, from 60% to 72%; at the same time the brightness decreases slightly, from magnitude −4.2 to −4.1. On 13 September, Venus passes 2.5° south of M44 (Praesepe or Beehive Cluster).

EARTH reaches its second equinox of the year on 22 September. The southern hemisphere greets the arrival of spring whilst the north awaits the onset of autumn.

MARS is again occulted by the Moon this month. On 6 September, observers in central South America, the central Atlantic Ocean region and north western Africa may see the waning gibbous Moon pass in front of the bright red planet from around 02:30 UT. Mars reaches a stationary point three days later and begins retrograde motion, a sure sign that opposition is not far away. Found in Pisces, Mars increases in brightness from magnitude −1.8 at the beginning of the month to −2.5 by the end. The planet now rises in early evening and is visible for much of the night.

JUPITER resumes direct motion on 13 September. The waxing gibbous Moon passes 1.6° south of the planet on 25 September. The brightest (non-lunar) object in Sagittarius at magnitude −2.5, Jupiter is already aloft at sunset. The best place to observe Jupiter is from southern latitudes where it doesn't set until midnight or after. However, by the end of the month, viewers in northern temperate regions lose the brilliant gas giant in late evening as the ecliptic through Sagittarius is close to the southern horizon at this time of year.

SATURN is found 2.3° north of the Moon on 25 September. Four days later it ends retrograde and returns to direct motion, heading back toward Capricornus. The rings have been opening up slightly since May and reach a maximum of 22.8° on the last day of the month before starting to close up again. Saturn is in Sagittarius and is visible in the evening, not setting until around or just after midnight. Southern hemisphere observers have the best views of this bright object, now dimming slightly from magnitude +0.2 to +0.4 over the course of the month.

URANUS is approaching opposition next month and appears in the east earlier every night. Now rising in the early evening, the best views of this faint planet are after midnight.

NEPTUNE is at opposition on 11 September, shining its brightest at magnitude +7.8. It rises around sunset and is visible all night. Aim your telescope toward Aquarius to spot this 2.5 arc-second-diameter object, now a mere 28.9 AU away from Earth. The accompanying finder chart shows the position of Neptune throughout 2020. Background stars are shown to magnitude 10.

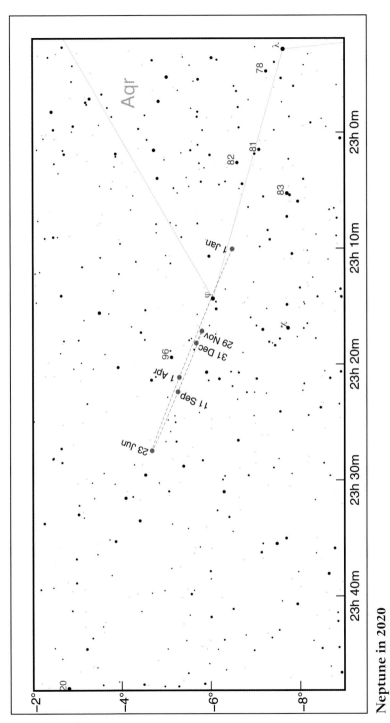

Neptune in 2020

A Closer Look at Indus

Brian Jones

Situated immediately to the south west of the constellation Grus, this tiny group of stars was first depicted on a celestial globe made by the Dutch cartographer Petrus Plancius in 1598. It first featured on star charts a few years later, making an appearance in the *Uranometria* star atlas published by German astronomer Johann Bayer in 1603.

The stars forming Indus were introduced by the Dutch navigators and explorers Pieter Dirkszoon Keyser and Frederick de Houtman following their expedition to the East Indies in the 1590s. Keyser had been asked by Plancius to chart the southern hemisphere stars during the voyage and, although Keyser died in 1596, the observations made were passed on to Plancius when the expedition returned in 1597. The constellation represents a native Indian,

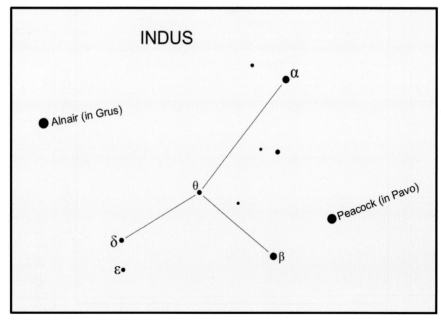

Indus. (Brian Jones/Garfield Blackmore)

although the country or region to which the character was indigenous is not known, Keyser and Houtman having visited many regions during their travels.

Indus is visible in its entirety from latitudes south of 15°N. None of the stars forming Indus are particularly bright and none are named. The neighbouring stars Alnair in Grus (located to the north east of Indus) and Peacock in Pavo (found immediately to the south west) are shown here as a guide to locating this faint group.

The brightest star in Indus is the orange giant Alpha (α) Indi, the magnitude 3.11 glow of which reaches us from a distance of around 100 light years. Shining from a distance of a little over 600 light years, magnitude 3.67 Beta (β) Indi is another orange giant. Delta (δ) Indi lies at the eastern end of Indus, the magnitude 4.40 glow of this yellowish-white star having set off towards us around 190 years ago. Closer examination will reveal that Delta Indi is actually a binary star with components of magnitudes 4.80 and 5.96 orbiting each other over a period of 12.2 years.

The somewhat undistinguished form of Indus is completed by magnitude 4.39 Theta (θ) Indi which shines from a distance of 100 light years. Theta is a double star, described by Ernst Johannes Hartung in his *Astronomical Objects for Southern Telescopes* as a "striking pair" comprising a pale-yellow primary and a reddish companion. The magnitude 4.6 and 7.2 components of Theta Indi are easy to separate and present themselves as an excellent target for small telescopes.

Another star of particular note is the magnitude 4.69 Epsilon (ε) Indi. Representing one of the four arrows held in the left hand of the Indian in the star atlas *Uranographia*, published in 1801 by German astronomer Johann Elert Bode, this orange dwarf lies at a distance of just 11.8 light years, making it one of our closest neighbours in space.

Although Indus is not the brightest of the constellations it should be reasonably straightforward to locate, situated as it is between the two leading stars of the neighbouring Grus and Pavo. Its original conception and presence on star charts is perhaps testament to the considerably darker, pre-light-pollution skies of the southern hemisphere under which Pieter Dirkszoon Keyser and Frederick de Houtman set sail during the late-16th century.

October

Full Moon: 1 October
New Moon: 16 October
Full Moon: 31 October

MERCURY reaches greatest elongation east (26°) on the first day of October. It enters retrograde motion on 14 October and undergoes inferior conjunction 11 days later, ending an excellent apparition for southern hemisphere observers. As is always the case, Mercury dims dramatically to sixth magnitude as it approaches inferior conjunction before brightening to +2.0 by the end of the month. Look for Mercury above the eastern horizon before sunrise by the end of October.

VENUS is found in the east before dawn, outshining every star and planet in the night sky. It is only 0.1° south of Regulus, the lucida of Leo, on the second day of the month. Perihelion occurs on 30 October. Northern latitudes are still favoured for viewing this bright −4.0 magnitude object.

EARTH is illuminated by two Full Moons this month. The first one, commonly called the Harvest Moon as it is the Full Moon nearest to the northern hemisphere autumnal equinox, occurs on the first day of the month whilst the second arrives on the final day. The second Full Moon is also this year's Micro Moon, the most distant and thus smallest Full Moon of 2020 (compare Super Moon in April). Furthermore, a second Full Moon in a calendar month is popularly known as a Blue Moon.

MARS has a busy month, starting with a lunar occultation on 3 October which is visible from the southern tip of South America and Antarctica. This event begins at approximately 02:00 UT. Mars reaches its minimum distance from Earth (only 0.415 AU) on 6 October, a week before opposition when it is at its brightest, shining at magnitude −2.6. The waxing gibbous Moon pays a second visit on 29 October when it passes 3° south of the red planet. Mars is in Pisces and is visible all night although it is best seen from southern latitudes.

JUPITER reaches east quadrature on 11 October. Astrophotographers will see the shadows of the planet and its major satellites cast slightly off to one side, leading to some interesting visual effects. The waxing crescent Moon comes to within 2° of Jupiter on 22 October. Jupiter is already above the horizon at sunset, shining at magnitude −2.3 in Sagittarius. It vanishes in the west by mid-evening for planet watchers in northern temperate zones but remains aloft until nearly midnight for those observing it from the southern hemisphere.

SATURN reaches east quadrature one week after Jupiter which makes this an excellent time to observe the planet, its rings and satellites through a telescope, especially since the very young Moon will not add to the brightness of the sky. The waxing crescent Moon does a fly by on 23 October when it passes 2.6° south of the planet. Found in Sagittarius after sunset, Saturn is at magnitude +0.4 and is high above the horizon as viewed from southern latitudes. Observers in northern temperate latitudes will struggle to see the ringed planet near the southern horizon.

URANUS reaches opposition on the last day of the month when it comes to within 18.8 AU of Earth. Uranus is only magnitude +5.7 and the Full Moon that night will flood the sky with light so binoculars or a small telescope will be required to spot this object which is just 3.7 arc-seconds in apparent diameter. The green ice giant rises around sunset and is visible all night in Aries. The accompanying finder chart shows the position of Uranus throughout 2020. Background stars are shown to magnitude 8.

NEPTUNE continues its post-opposition retrograde path across Aquarius. This eighth-magnitude planet is visible in the evening sky and is best viewed (with optical aids, of course) from the southern hemisphere where the ecliptic rises high into the sky.

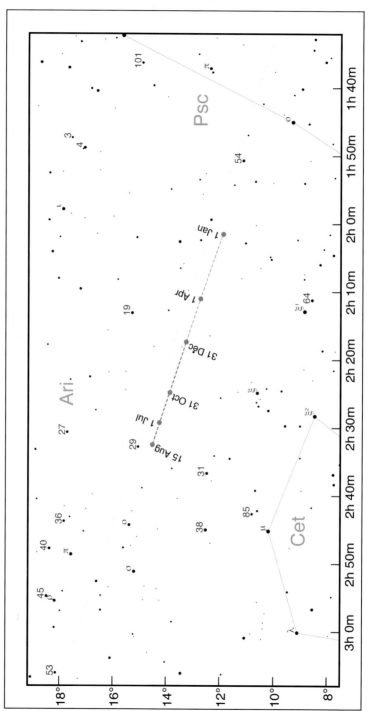

Uranus in 2020

Airbursts:
A Spectacular Astronomical Phenomenon

Susan Stubbs

Our planet occupies a place in our solar system that is filled with dust and debris. Around 100,000 tonnes of spacedust and interplanetary particles enter our atmosphere each year, this being comprised mostly of various dusts and small grains present in the early solar system. Most particles are smaller than a grain of sand and so burn up in our atmosphere; these may be visible as 'shooting stars'. Meteors brighter than magnitude −4 are called bolides, or fireballs, whilst those of magnitude −13 or more are superbolides. On rare occasions an object of sufficient size will survive its journey through our atmosphere and explode above the surface of the Earth where it has the potential to cause significant damage and casualties. These larger bodies, or debris from them, are known as Near Earth Objects (NEOs), and have orbital paths that bring them close the Earth's orbit. There are over 19,000 near-Earth asteroids and over a hundred near-Earth comets that are large enough to be tracked in space and which have been detected in the Earth's neighbourhood. An object as small as 20m can cause sufficient heat and shock wave damage to devastate local environments and populations.

Large superbolides with explosive energy similar to that of the weapons of war, known as airbursts, are usually due to objects greater than 1 metre diameter; such objects enter the Earth's atmosphere several times a year, although most vaporise or disintegrate completely in the upper layers. The entry speed and angle, composition and density of the object all influence whether it breaks up, explodes as an airburst or survives to form a crater on the surface. The best known airburst of recent times is that which appeared in the skies over Chelyabinsk in west-central Russia in February 2013. Prior to that the largest witnessed event took place over Tunguska in June 1908. Many more around the globe have been inferred by the witnessing of fireballs and later ground damage, through finding widespread meteoritic fragments in the absence of a crater or, in recent years, by infrasound data from the Monitoring

Flattened trees at Tunguska site, imaged during the expedition led by Leonid Kulik in 1929. (Vokrug Sveta / Евгений Леонидович Кринов / Wikimedia Commons)

System of the Comprehensive Nuclear Test Ban Treaty (set up to monitor illicit nuclear tests). The 1490 Ch'ing Yang event, for which documentation is scarce, is recorded by Chinese scholars as having caused 10,000 deaths. Past collisions have certainly played a significant role in shaping the geology and biology of the Earth throughout its history.

The Tunguska explosion occurred over a remote area of Siberia on 30 June 1908. It flattened around 2,000 square km of forest and is the largest impact event on Earth in recorded history. There were no known human casualties, largely due to the remoteness of the location, but there were eyewitnesses who described a bright light moving across the sky followed by an explosion and a shockwave that knocked people off their feet. Windows were broken hundreds of kilometres away and the explosion detected at seismic stations around the world. It is estimated to have been equivalent to 15,000 kilotonnes of TNT, 1000 times greater than the Hiroshima atomic bomb, and the meteoroid itself to have been anything between 60 metres and 190 metres in diameter. The first recorded scientific investigation arrived there in 1929, headed by the mineralologist Leonid Kulik. They found no crater, but widespread devastation and soil mineral deposits suggestive of an airburst.

The Chelyabinsk meteor was a superbolide resulting from a 20 metre, 13,000 tonne, asteroid remnant that entered the atmosphere over Russia at 09:20 local time on 15 February 2013. It is likely to have been rubble from the near-Earth asteroid 1999 NC43, which crosses the Earth's orbit every few years but which is not in itself a threat to us. The meteor exploded at a height of around 30km, due to its shallow angle of entry and high velocity, resulting in a bright flash and shockwave seen and felt by thousands. Due to modern technology this event was photographed and videoed by many, and so much more is known about this airburst than any other. The explosion was equivalent to between 400 and 500 kilotons of TNT, causing widespread damage to buildings and around 1,500 people requiring medical treatment for injuries. It was not detected prior to entry by NEO monitoring due to its radiant being close to the Sun. The Chelyabinsk meteoroid yielded over 1,000 kg of chondritic meteorites, with the largest fragment weighing 540g being found 70 km from the main blast site.

An image of the flare and falling debris from the Chelyabinsk meteoroid. (Константин Кудинов/ Wikimedia Commons)

What is the risk to us from airbursts, and impacts from meteors and asteroids? NEOs have become of increasing interest in recent years because of the potential danger they may pose, and monitoring by organisations such as NASA aims to identify those that are a threat. There are surprisingly few reports of injuries or fatalities from space debris, however. Those events recorded include that of a cow being killed in Venezuela in 1972, a car damaged in the United States in 1992, and a house on the outskirts of Paris being hit by a meteorite in 2011. The Chinese report of 10,000 deaths in 1490 has no verification.

Around 2,000 asteroids have been detected that will pass to within 8 million km of us (and are thus defined as Potentially Hazardous Asteroids). A number of these are over 1 km in diameter and are therefore potentially globally catastrophic. There are many thousands more smaller objects with diameters of between 5 and 10 metres that could cause an airburst releasing energy equivalent to 15,000 tons of TNT and a resulting shock wave of up to 10,000K in temperature. These enter the atmosphere at a rate of approximately once or twice yearly, although we know of no large object likely to strike us in the next few hundred years. The information from the Chelyabinsk event has contributed enormously to scientific knowledge about airbursts. For the first time scientists can link the well-described damage to the known impact energy, and estimate the range of probabilities of events and the likely consequences to us. It has also led some to wonder if the current estimates of the probability of impact events are too low. We are powerfully reminded of the vulnerability of our planet when such events occur.

November

New Moon: 15 November
Full Moon: 30 November

MERCURY begins its final morning apparition of the year. After enjoying an excellent evening apparition over the past few months, observers in the southern hemisphere now get their worst morning show. Conversely, planet watchers in northern temperate latitudes have their best dawn apparition of the year. Mercury steadily brightens throughout the month, beginning at magnitude +1.5 and ending at −0.8. The planet reaches its final perihelion of 2020 on 2 November and returns to direct motion the following day. Greatest elongation west (19°) occurs on 10 November and the very old crescent Moon passes 1.7° north of Mercury three days later.

Morning Apparition of Mercury
25 October to 20 December

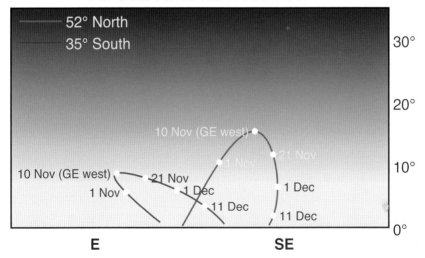

VENUS is still well above the eastern horizon before sunrise but it is getting lower every day. Northern and equatorial early risers have much the best views of the morning star. When viewed through a telescope, Venus appears in its waxing gibbous phase, with the illuminated fraction increasing from 81% to 89% and the magnitude holding steady at −4.0. The waning crescent Moon is just 3.1° north of Venus on 12 November.

EARTH experiences the fourth and final lunar eclipse of 2020 on the last day of the month when the Full Moon passes through our planet's penumbra. More information on all of this year's eclipses may be found in the section *Eclipses in 2020*.

MARS is now past opposition, getting more distant from Earth and starting to dim. It drops an entire magnitude this month, starting at a brilliant −2.1 and ending November at a still respectable −1.1. Mars ends its retrograde motion in Pisces on 15 November. It is already up when the sky grows dark and sets in the early morning hours, later for northern latitudes than for southern.

JUPITER, at magnitude −2.1 in Sagittarius, is best viewed from southern latitudes where it doesn't set until late in the evening. Observers in northern temperate zones lose the bright planet much earlier as it is already low to the horizon. The waxing crescent Moon passes 2.5° south of Jupiter on 19 November. Saturn is in close proximity which will make a pretty picture in the west as the sky darkens.

SATURN is, like Jupiter, an evening sky object in Sagittarius. At magnitude +0.5, it is slowly dimming as the rings close and the planet draws away from Earth. Saturn sets in the early evening as seen from northern temperate latitudes but remains above the horizon a little longer for those observing from the southern hemisphere. The waxing crescent Moon comes within 3° of Saturn on 19 November.

URANUS reached opposition on the last day of October so it is rising around sunset and setting around sunrise for all locations. Choose a moonless night around the middle of the month to see the sixth-magnitude planet in the faint constellation of Aries.

NEPTUNE concludes its retrograde loop on 29 November and returns to direct motion across Aquarius. The faint ice giant is already above the horizon as darkness falls, setting in the early morning hours.

Toads, Earthworms and Slugs Among the Stars: The Curious Constellations of Dr John Hill

David Harper

For as long as humans have gazed up at the night sky, they have grouped the stars into shapes representing mythic heroes such as Hercules and Perseus, powerful animals like lions, bears and bulls, and creatures of fantasy such as Pegasus the winged horse. The 88 constellations that are officially recognised by the International Astronomical Union include more mundane objects such as tables, cups and flies. If Dr John Hill had had his wish, there would also be constellations depicting a toad, a slug, an earthworm and a leech, among others.

John Hill – the date of the portrait is unknown. (Wikimedia Commons/ Makers of British Botany)

Hill was born in 1714 in Peterborough, Northamptonshire. We know little about his early years, but by 1738, having gained some medical education, he was working in London as an apothecary. He wrote extensively on many subjects, notably botany. Between 1759 and his death in November 1775, he published a 26-volume treatise on plants called *The Vegetable System*.

His wide-ranging interests also included astronomy and mineralogy. He knew many prominent members of the Royal Society, and aspired keenly to join their ranks. Unfortunately, he also penned a number of scathing critiques of what he judged to be the failings of the Society. He was particularly critical of the standard of papers which the Society published in its journal *Philosophical Transactions*, citing titles such as *An Account of a Mer-Man* and *On Demons in Lead-Mines*. Unsurprisingly, these attacks won him no friends within the Society, and all of his attempts to become a Fellow failed.

Hill's lasting contribution to astronomy was a volume published in 1754 under the title *Urania, or a Compleat View of the Heavens*. This took the form of a dictionary of more than 600 pages. In the preface, Hill explained that it was intended as the first volume in a series which would eventually encompass geology, metallurgy, botany, biology and mathematics. Subsequent volumes never materialised, so we must be grateful that Hill began with astronomy.

Most of the entries are the names of planets, stars and constellations, with explanations of their origins. In a sense, *Urania* may be seen as a forerunner of Richard Hinckley Allen's *Star Names: Their Lore and Meaning*. Other entries explained astronomical concepts and technical terms. The level of detail is often impressive: the entry for Mars covers nine pages and contains a lengthy discussion of the history of telescopic observation of the planet, including Hill's own observations.

Urania was published a year before Samuel Johnson's great dictionary of the English language, and both dictionaries include pithy and humorous comments which convey the character of their respective authors. Hill's entry for **almantar** begins thus:

A name by which the astrological writers, when they have a mind to be more than ordinarily obscure, have called what they generally express by the term Aspect.

Astronomers do not escape his barbs. The adjacent entry for **almicantarahs** complains that "astronomers are too fond of hard words", whilst in the entry for Monoceros (the Unicorn), Hill chides the 17th century astronomer Johannes Hevelius:

When there were so many real animals for Hevelius to have chosen amongst, it was very idle in him to have fixed on an imaginary one.

Hill had a particular detestation of people who deliberately use obscure or "hard" words. The entry for **Moloch Baal** is typical of many:

A name by which those, who are fond of introducing hard names on every occasion, call the planet Jupiter; it is a term that we find indeed used by very early writers, but that is no reason why late ones should use it.

Amongst the hundreds of entries, fifteen stand out from the rest. These are new constellations, invented by Hill himself from stars in the "gaps" between established groups. He introduces each one with uncharacteristically tentative wording. For example, the entry for **Bufo**, the toad, begins in this way:

A constellation offered to the astronomical world, and composed of a number of unformed stars near the sign Libra.

He goes on to explain how the stars depict the animal, noting:

The stars in the constellation Bufo are fifteen, and some of them are remarkably bright and considerable; in the head there are only two, one is near the extremity or mouth, and this is a small one; the other, which may be called the Toad's Eye, is a very fine one of second magnitude.

The Toad's Eye is Sigma (σ) Librae, a red giant which is a semi-regular variable. Its magnitude ranges between 3.2 and 3.5. Most of the other stars of Bufo are in the eastern end of Hydra, and all are much fainter than the Toad's Eye.

Hill appears to have had a fondness for invertebrates. He also proposed the constellations of Aranea the spider (formed from stars near Virgo), Lumbricus the earthworm (between Cancer and Gemini), Limax the slug (beneath Orion's foot) and Hirudo the leech (above Orion's head). His other candidates represented a tortoise, a seahorse, an eel, four species of shellfish, the scarab beetle, the stargazer fish (which rejoices in the Latin name Uranoscopus) and, somewhat bizarrely, the pangolin or scaly anteater.

Urania is the only work which lists Hill's constellations. Allen does not mention them, nor does he include Hill in his bibliography, so we must assume that he was unaware of his industrious and inventive predecessor. Most of Hill's new constellations were formed from a handful of faint stars which bore scant resemblance to the creature that Hill imagined them to represent. Ironically, Hill himself includes this barbed comment in the entry for **Apis** the bee, a name once given to the modern constellation Musca (the Fly):

This little constellation is very aptly represented in the drawings of the heavens, but it is not so well expressed by the stars that are comprised in

The whitemargin stargazer (Uranoscopus sulphureus), the creature in honour of which John Hill named his constellation the stargazer fish, spends most of its life buried in sand or mud, with only its eyes showing, and attracts prey using a lure. (Wikimedia Commons/U.S. National Park Service/Richard C. Wass)

it; all of it that the authors of it meant by the figure, seems to have been, that, having a very small cluster of stars to describe, they chose as little a figure as they well could to place them in.

Notwithstanding Hill's utter failure to persuade astronomers to add toads, slugs and earthworms to the heavens, his *Urania* was a remarkable accomplishment for its time, and it remains a fascinating and entertaining example of 18th century astronomical writing. Happily, it has been digitised by Google Books and **archive.org** and there is also a facsimile printed edition, although the latter lacks most of the star charts which depict Hill's constellations.

December

New Moon: 14 December
Full Moon: 30 December

MERCURY continues to decline in altitude above the eastern horizon, vanishing from the dawn sky before the 20 December superior conjunction, only to reappear in the west at sunset at the end of the month. It remains fairly bright, varying between magnitudes −0.8 and −1.3.

VENUS is just 0.8° south of the waning crescent Moon on 12 December. Observers in north eastern Asia and northern Alaska may see the planet occulted by our satellite starting about 19:00 UT. The morning star is slowly descending toward the horizon as it closes in on the Sun and superior conjunction next year. The illuminated part of its disk increases from 89% to 94% but because it is getting farther from Earth, its apparent diameter is decreasing, so the magnitude of Venus stays constant at −4.0 this month.

EARTH experiences a total solar eclipse on 14 December when the New Moon briefly obscures the face of the Sun. See the section *Eclipses in 2020* for more details. Solstice is one week later. This brings the longest day of the year for the southern hemisphere and the longest night for the north.

MARS passes through its ascending node on the second day of the month, returning to the north side of the ecliptic. Except for the Moon, it is the brightest object in the faint constellation of Pisces, shining at magnitude −1.1 on 1 December and ending the year at −0.2. Mars is well-placed for evening observations for planet chasers in northern temperate latitudes where it doesn't set until well after midnight. Southern hemisphere viewers, however, must look for the red planet soon after the skies darken as Mars sets before midnight by the end of the year.

JUPITER is busy mid-month, with a fly-by of the ninth-magnitude globular cluster M75 around 16 December, a close encounter with the waxing crescent Moon on 17 December and crossing the border into Capricornus on 18 December. The main event, however, is the Jupiter–Saturn conjunction on 21 December when Jupiter (magnitude −2.0) passes 0.1° south of Saturn (magnitude +0.6). The two objects are 30° away from Sun so the event should be visible in the west after dark.

SATURN is at solar conjunction late next month so it is getting more difficult to see in the western twilight. Saturn returns to Capricornus on 15 December (it retrograded out of the constellation in July) and the very young crescent Moon passes south of the ringed planet two days later. On 21 December, Jupiter and Saturn are in conjunction, both in ecliptic longitude and right ascension, the first time this has happened since May 2000. The two gas giants come to within 0.1° of each other. More information relating to this event can be found in David Harper's article *Jupiter and Saturn – At Their Closest in Almost Four Centuries* elsewhere in this volume.

URANUS ends the year as it began, visible in the evening sky in Aries and undergoing retrograde motion. Northern planet watchers get the best views, with Uranus not setting until well after midnight.

NEPTUNE reaches east quadrature on 9 December. Look for it in Aquarius before it sets around midnight.

Jupiter and Saturn:
At Their Closest in Almost Four Centuries

David Harper

On the evening of 21 December 2020, Jupiter and Saturn will be just 0.1° apart. This is their closest conjunction since 16 July 1623, when they were separated by 0.09°. That event could not have been observed, as the two planets were only 12° from the Sun. This year, we are more fortunate. The planets are 30° east of

The Star of Bethlehem appears twice in this painting of the Adoration of the Magi by the 14th-century/15th century French artist Jean Fouquet. A conjunction of Jupiter and Saturn is a popular explanation for the Star of Bethlehem. (Jean Fouquet/Wikimedia Commons)

the Sun on 21 December, and they will be visible in the evening sky. The best time to look is 30 minutes to an hour after sunset, when the sky will be dark enough to see first-magnitude Saturn, but the planets will still be at least 10° above the horizon.

Jupiter and Saturn are separated by less than a degree from 12 December until 30 December, although by the end of the month, they will be only 23° from the Sun, making it increasingly difficult to see the fainter Saturn in the twilight sky.

Conjunctions of Jupiter and Saturn occur at intervals of roughly twenty years. The last one was in May 2000, but as in 1623, the planets were too close to the Sun for the event to be visible. By contrast, in 1980 and 1981, astronomers were treated to a triple conjunction. Jupiter and Saturn were barely a degree apart on 31 December 1980 and again on 4 March and 24 July 1981. All three conjunctions were easily seen, as both planets were at opposition in late March 1981.

Only one conjunction in seven is a triple conjunction, because this phenomenon relies upon Jupiter's retrograde motion around opposition to carry it past Saturn repeatedly. Of the 1,511 conjunctions between the years –13,000 and +17,000, 1301 are single events (like the conjunction of 2020) and 210 are triples.

Nor do triple conjunctions occur regularly. Whilst the 20th century had two, in 1940/41 and 1980/81, there are none during the present century, or in the 22nd century. Astronomers must wait until the *Yearbook of Astronomy 2239* to read about the next such event.

The triple conjunction of the year –6 (7 BCE) is sometimes proposed as a

Johannes Kepler (1571–1630) was a German mathematician and astronomer who studied planetary orbits. He was the first person to propose the triple conjunction of Jupiter and Saturn in the year –6 (7 BCE) as an explanation for the Star of Bethlehem. Portrait of Johannes Kepler from a lost original of 1610 in the Benedictine monastery in Kremsmünster. (Wikimedia Commons)

candidate for the Star of Bethlehem which is mentioned in the Gospel of St. Matthew. This idea was originally put forward by the German astronomer Johannes Kepler, who first calculated that a triple conjunction occurred in that year. It was re-popularised in the 1970s by Professor David Hughes of Sheffield University, who suggested that the Magi were astrologers who read astrological significance into the event. Other astronomical explanations include a supernova, a bright comet, or even a triple conjunction between Jupiter and Regulus followed by a conjunction of Jupiter with Venus. Each explanation has its strengths and weaknesses, and since historical records are silent on the subject, astronomers will continue to speculate on the correct answer for centuries to come.

Comets in 2020

Neil Norman

For the comet enthusiast, 2020 is promising to be something of a barren year. In total, 71 comets will approach perihelion in 2020, and of these, no fewer than 61 are periodic comets belonging to the Jupiter family and having orbital periods of 20 years or less. However, these comets are a double-edged sword for observers. On the one hand they appear frequently and can become old friends, but on the other, their repeated perihelion passages over thousands of years has resulted in them expending much of their volatile materials, leaving them as rather dim objects.

These comets are very much left to the amateur domain because they are of no scientific value to the professionals, unless they make close approaches to Earth or undergo outbursts, as was the case with Comet 17P/Holmes in 2007. Discovered by English amateur astronomer Edwin Alfred Holmes on 6 November 1892 and with an orbital period of 6.9 years, this comet is normally rather faint, but became quite a notable object during its return in October 2007 when it brightened temporarily to around magnitude 3 and became visible to the naked eye.

It is worth remembering that comets are being discovered on a regular basis, and that the next 'great' comet could be discovered at any time! You can follow the latest discoveries by checking out the British Astronomical Association Comet Section page **www.ast.cam.ac.uk/~jds** where you can read the prospects for any newly discovered comets.

Best Prospects For 2020

You will need some form of optical aid, such as binoculars or a small telescope, in order to observe the comets detailed below. However, predicting comets is not a precise science and they can either exceed expectations or not live up to predictions. Therefore, regular monitoring of the BAA website is strongly advised.

C/2017 T2 (PanSTARRS)

This comet was discovered on 2 October 2017 by the 1.8m Panoramic Survey Telescope and Rapid Response System (Pan-STARRS) Ritchey–Chrétien Telescope at Haleakalā Observatory on the island of Maui, Hawaii. It was first spotted at magnitude 19.9, when situated at a distance of 9.25 AU from the Sun (around the orbital distance of Saturn) in the constellation Eridanus. The comet is well placed for northern hemisphere observers throughout 2020, during its approach to perihelion on 6 May when it will be 1.61 AU from the Sun and attain 8th magnitude. Comet C/2017 T2 (PanSTARRS) will remain below magnitude 10 until around 7 August.

The PanSTARRS1 Observatory at Haleakalā, Maui, seen here just before sunrise. (Rob Ratkowski/ Pan-STARRS1 Surveys and Archive/panstarrs.stsci.edu)

DATE	R.A.	DEC	MAG	CONSTELLATION
1 Jan 2020	03 26 58	+55 16 20	9.5	Perseus
1 Feb 2020	02 17 37	+58 04 04	9.2	Perseus
1 Mar 2020	02 10 34	+61 25 57	8.9	Cassiopeia
1 Apr 2020	03 03 29	+68 51 45	8.5	Cassiopeia
1 May 2020	06 19 36	+76 15 19	8.3	Camelopardalis
1 Jun 2020	10 54 59	+64 45 50	8.3	Ursa Major
1 Jul 2020	12 31 35	+41 32 13	8.8	Canes Venatici
1 Aug 2020	13 27 50	+19 28 12	9.7	Coma Berenices
1 Sep 2020	14 13 37	+03 24 05	10.7	Virgo

88P/Howell

This comet was discovered as a 15th magnitude object in Cetus on photographic plates dating to 29 August 1981 taken with the 46cm Palomar Schmidt telescope by American astronomer Ellen Suzanne Howell. 88P/Howell is a Jupiter family comet with an orbital period of 5.48 years, and will undergo its next perihelion on 26 September 2020. Generally speaking, 88P/Howell is a very diffuse object, making it a good potential target for binoculars.

Comet 88P/Howell imaged from Siding Spring, Australia on 19 March 2015. (José J. Chambó)

DATE	R.A.	DEC	MAG	CONSTELLATION
1 May 2020	12 58 25	−01 47 08	11.0	Virgo
15 May 2020	12 46 29	−01 22 58	10.7	Virgo
1 Jun 2020	12 40 41	−01 56 24	10.2	Virgo
15 Jun 2020	12 44 19	−03 17 20	10.0	Virgo

29P/Schwassmann-Wachmann

Discovered at the Hamburg Observatory by German astronomers Arnold Schwassmann and Arno Arthur Wachmann on photographs exposed on 15 November 1927, this comet is in a class called Centaurs group, of which 80 are known. These bodies are believed to have escaped the Kuiper belt which is located just beyond Neptune at between 30–50 AUs. 29P/Schwassmann-Wachmann has an almost-circular orbit with a period of 14.65 years. The comet has been in its current orbit for 40,000 years or so and will eventually be put either into a long period orbit or expelled from the Solar System by interactions with Jupiter. This object is known for its outbursting behaviour where it can rise from its usual 13th magnitude to magnitude 9 in the course of a day. All reliable observations of 29P/Schwassmann-Wachmann are welcomed, and can be submitted to the BAA.

Comet 29P Schwassmann-Wachmann imaged on 7 September 2017 from Harlingten Observatory, Spain by Tony Angel and Caisey Harlingten. (Tony Angel and Caisey Harlingten)

DATE	R.A.	DEC	MAG	CONSTELLATION
15 Jan 2020	00 33 26	+13 02 18	13 – may vary	Pisces
1 Feb 2020	00 41 41	+13 34 27	13 – may vary	Pisces
15 Feb 2020	00 49 59	+14 12 51	13 – may vary	Pisces
1 Mar 2020	01 00 04	+15 03 42	13 – may vary	Pisces
15 Mar 2020	01 10 18	+15 58 12	13 – may vary	Pisces
1 Aug 2020	02 47 01	+25 47 09	13 – may vary	Aries
15 Aug 2020	02 50 59	+26 27 11	13 – may vary	Aries
1 Sep 2020	02 52 54	+27 04 51	13 – may vary	Aries
15 Sep 2020	02 51 53	+27 24 48	13 – may vary	Aries
1 Oct 2020	02 47 55	+27 32 45	13 – may vary	Aries
15 Oct 2020	02 42 24	+27 25 19	13 – may vary	Aries
1 Nov 2020	02 34 08	+26 58 30	13 – may vary	Aries
15 Nov 2020	02 27 09	+26 24 42	13 – may vary	Aries
1 Dec 2020	02 20 16	+25 39 27	13 – may vary	Aries
15 Dec 2020	02 16 06	+25 00 25	13 – may vary	Aries

In summary, although 2020 is destined to be somewhat disappointing for many comet observers, droughts like this do occur. The next significant comet could be discovered at any time, and a good source for the latest news is the BAA Comet site.

Minor Planets in 2020

Neil Norman

Minor planets is just another term for asteroids, the varying sized pieces of rock left over from the formation of the Solar System around 4.6 billion years ago. Millions of them exist, and to date some 780,000 have been seen and documented, with around 450,000 having received permanent designations after being observed on two or more occasions.

Most asteroids travel around the Sun in the main asteroid belt, which is located between the orbits of Mars and Jupiter. However, some asteroids have more elliptical orbits which allow them to interact with major planets, including the Earth. In all, there are around 2,000 which can approach to within a close distance of our planet. These objects are referred to as Potentially Hazardous Asteroids (PHAs). To qualify as a PHA these objects must have the capability to pass within 8 million km (5 million miles) of Earth and to be over 100 metres across.

Objects of this size could pose a serious threat if on a collision course with Earth. It is estimated that several thousand exist with diameters of over 100 metres, and with around 150 of these being over a kilometre across. A large number of smaller asteroids, measuring anything between just a few meters in diameter to several tens of meters wide, pass close to our planet on a regular basis, with considerable numbers of smaller ones entering the Earth's atmosphere every day, burning up harmlessly as meteors.

Those observers with a particular interest in following these objects should go to the home page of the Minor Planet Center, who's job it is to keep track of these objects and determine orbits for them. This page can be accessed by going to **www.minorplanetcenter.net** where you will find a table of newly discovered minor planets and Near Earth Objects (NEOs). At the top of the page is a search box that you can use to locate information on any object that you are interested in, and from this you can obtain ephemeredes of the chosen subject. The Minor Planet Center site is the one that all dedicated asteroid observers should consult on a regular basis.

2 Pallas

Pallas has a diameter of 512 km (318 miles) and was the second asteroid to be discovered when first spotted by the German astronomer Heinrich Wilhelm Matthias Olbers on 28 March 1802. He named the object after the Greek goddess of wisdom and warfare Pallas Athena, an

Lithograph portrait of the German astronomer Heinrich Wilhelm Matthias Olbers by the German portrait painter and lithographer Rudolf Suhrlandt. (Wikimedia Commons)

alternative name for the goddess Athena. Pallas's orbit is highly eccentric and its path around the Sun steeply inclined to the main plane of the asteroid belt, rendering it fairly inaccessible to spacecraft.

DATE	R.A.	DEC	MAG	CONSTELLATION
1 Mar 2020	18 53 53	+ 07 58 17	10.4	Aquila
1 Apr 2020	19 22 39	+ 12 11 39	10.3	Aquila
1 May 2020	19 37 00	+ 16 47 15	10.1	Sagitta
1 Jun 2020	19 33 49	+ 20 44 52	9.8	Vulpecula
1 Jul 2020	19 14 25	+ 21 53 34	9.6	Vulpecula
1 Aug 2020	18 05 35	+ 18 57 45	9.6	Hercules
1 Sep 2020	18 39 57	+ 13 11 38	9.9	Hercules
1 Oct 2020	18 47 41	+ 07 25 56	10.3	Aquila
1 Nov 2020	19 10 52	+ 02 53 11	10.5	Aquila
1 Dec 2020	19 42 55	+ 00 17 42	10.6	Aquila

3 Juno

Discovered by German astronomer Karl Ludwig Harding on 1 September 1804, and with a mean diameter of 233 km (145 miles), Juno is the second-largest of the stony (S-type) asteroids, exceeded in size only by 15 Eunomia (see below).

DATE	R.A.	DEC	MAG	CONSTELLATION
1 Feb 2020	13 24 31	− 05 04 12	10.4	Virgo
1 Mar 2020	13 23 07	− 02 40 15	10.0	Virgo
1 Apr 2020	13 03 20	+ 01 35 10	9.5	Virgo
1 May 2020	12 42 03	+ 04 53 39	10.0	Virgo
1 Jun 2020	12 35 04	+ 05 40 12	10.6	Virgo

4 Vesta

Another discovery by Heinrich Wilhelm Matthias Olbers, who first spotted it on 29 March 1807, Vesta is one of the largest asteroids, with a diameter of 525 km (326 miles) and an orbital period of 3.63 years. This is the second most massive object in the main asteroid belt, and holds the distinction of being the brightest minor planet visible from Earth.

DATE	R.A.	DEC	MAG	CONSTELLATION
15 Mar 2020	03 43 05	+ 16 48 45	8.3	Taurus
1 Apr 2020	04 07 49	+ 18 36 44	8.4	Taurus
1 Oct 2020	09 37 15	+ 16 20 11	8.3	Leo
1 Nov 2020	10 25 44	+ 13 07 44	8.0	Leo
1 Dec 2020	11 05 13	+ 10 39 00	7.8	Leo

This image of the minor planet Vesta in Ophiuchus was taken on 15 August 2018 by Loren Ball, Decatur, northern Alabama. (Loren Ball)

6 Hebe

Discovered by the German amateur astronomer Karl Ludwig Hencke on 1 July 1847, Hebe is most likely to be the parent body of the H-type ordinary chondrites, the most common type of meteorite and which account for about 40% of all meteorites hitting Earth.

Hebe was one of two asteroids discovered by Karl Ludwig Hencke, the other being Astraea, which came to light on 8 December 1845. (Wikimedia Commons/The Sidereal Messenger)

DATE	R.A.	DEC	MAG	CONSTELLATION
15 Jan 2020	13 32 51	+ 01 55 07	11.0	Virgo
15 Feb 2020	13 46 20	+ 04 07 35	10.6	Virgo
15 Mar 2020	13 39 12	+ 08 12 59	10.0	Virgo
15 Apr 2020	13 14 58	+ 12 30 17	10.0	Virgo
15 May 2020	12 54 45	+ 13 44 21	10.2	Virgo
15 Jun 2020	12 50 59	+ 11 49 10	10.9	Virgo

8 Flora

On 18 October 1847 the English astronomer John Russell Hind discovered Flora, the name of this asteroid being proposed by John Frederick William Herschel in honour of the Roman goddess of flowers and gardens. With a mean diameter of 128 km (80 miles), Flora is the innermost large asteroid to orbit the Sun.

DATE	R.A.	DEC	MAG	CONSTELLATION
1 Aug 2020	02 12 33	+ 05 56 15	9.9	Cetus
1 Sep 2020	02 50 50	+ 06 50 10	9.3	Cetus
1 Oct 2020	03 02 42	+ 05 27 51	8.6	Cetus
1 Nov 2020	02 42 22	+ 03 13 26	8.0	Cetus
1 Dec 2020	02 18 48	+ 03 36 07	8.6	Cetus

18 Melpomene

Another of the asteroids discovered by John Russell Hind, Melpomene has a diameter of 170 km (105 miles) and first came to light on 24 June 1852. Named after the Muse of Tragedy in Greek mythology, Melpomene is an S-type asteroid that is composed of silicates and metals. Its orbital eccentricity is low (0.218) ensuring its path around the Sun is virtually circular and quite stable.

DATE	R.A.	DEC	MAG	CONSTELLATION
15 Sep 2020	07 48 41	+ 13 41 57	11.0	Gemini
15 Oct 2020	08 38 45	+ 10 51 47	10.9	Cancer
15 Nov 2020	09 15 27	+ 08 11 36	10.7	Cancer
15 Dec 2020	09 29 49	+ 07 13 11	10.3	Leo

15 Eunomia

Discovered on 29 July 1851 by the Italian astronomer Annibale De Gasparis, and named after the Greek goddess of law and legislation, Eunomia has a diameter of 357 km (222 miles) and is the largest of the stony asteroids, containing approximately 1% of the mass of the asteroid belt.

DATE	R.A.	DEC	MAG	CONSTELLATION
15 Jan 2020	22 53 27	+ 03 01 05	10.0	Pisces
1 Feb 2020	23 24 52	+ 06 02 06	10.0	Pisces
1 Sep 2020	07 05 37	+ 27 30 16	10.2	Gemini
1 Oct 2020	07 56 28	+ 24 50 02	10.1	Gemini
1 Nov 2020	08 33 24	+ 21 41 50	9.8	Cancer
1 Dec 2020	08 47 24	+ 19 08 33	9.4	Cancer

324 Bamberga

Discovered by Austrian astronomer Johann Palisa on 25 February 1892, and with a diameter of 229 km (142 miles), Bamberga is the fourteenth-largest, and tenth-brightest, asteroid in the asteroid belt. Bamberga is classified as a C-type asteroid, objects which are carbonaceous in composition and account for approximately 75% of known asteroids.

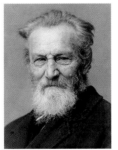

Johann Palisa was a prolific discoverer of asteroids, with over 120 discoveries to his name and all made by visual observation rather than photography. (Wikimedia Commons)

DATE	R.A.	DEC	MAG	CONSTELLATION
15 Jan 2020	12 32 09	− 09 42 11	13.0	Virgo
15 Feb 2020	12 26 40	− 10 55 32	12.6	Virgo
15 Mar 2020	12 06 19	− 10 13 44	12.0	Virgo
15 Apr 2020	11 41 20	− 08 08 55	12.3	Crater
15 May 2020	11 29 34	-06 32 27	12.0	Crater

Meteor Showers in 2020

Neil Norman

Whether you are a dedicated observer or just a casual sky gazer, on any given night of the year you can reasonably expect to see several meteors dash across the sky. Quite often these will be 'sporadic' meteors, that is to say they can appear at any time and from any direction. Sporadic meteors arise when a meteoroid – perhaps a particle from an asteroid or a piece of cometary debris orbiting the Sun – enters the Earth's atmosphere and burns up harmlessly high above our heads, leaving behind the streak of light we often refer to as a "shooting star".

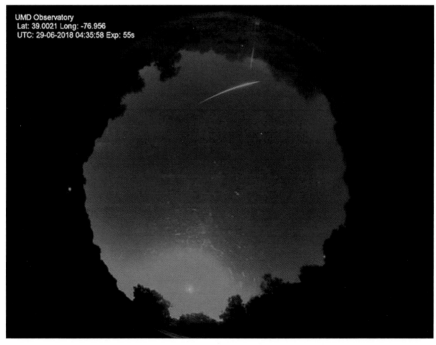

Sporadic meteor captured on 29 June 2018 by the Starlight Xpress Oculus camera with 180° lens at UMD Observatory, Maryland. (MASNO1, UMD Observatory)

The meteoroids in question are usually nothing more than pieces of space debris that the Earth encounters in its path as it travels along its orbit, and range in size from a few millimetres to a couple of centimetres in size. Meteoroids that are large enough to at least partially survive the passage through the atmosphere, and which reach the Earth's surface without disintegrating, are known as meteorites.

At certain times of the year the Earth encounters more organised streams of debris that produce meteors over a regular time span and which seem to emerge from the same point in the sky. These are known as meteor showers. These streams of debris follow the orbital paths of comets, and are the scattered remnants of comets that have made repeated passes through the inner solar system. The ascending and descending nodes of their orbits lie at or near the plane of the Earth's orbit around the Sun, the result of which is that at certain times of the year the Earth encounters and passes through a number of these swarms of particles.

The term 'shower' must not be taken too literally. Generally speaking, even the strongest annual showers will only produce one or two meteors a minute at best, this depending on what time of the evening or morning that you are observing. One must also take into account the lunar phase at the time, which may significantly influence the number of meteors that you see. For example, a full moon will probably wash out all but the brightest meteors.

The following is a table of the principle meteor showers for 2020 and includes the name of the shower; the period over which the shower is active; the ZHR (Zenith Hourly Rate); the parent object from which the meteors originate; the date of peak shower activity; and the constellation in which the radiant of the shower is located. Most of the information given is self-explanatory, but the Zenith Hourly Rate (ZHR) may need some elaborating.

The Zenith Hourly Rate is the number of meteors you may expect to see if the radiant (the point in the sky from where the meteors appear to emerge) is at the zenith (or overhead point) and if the observing conditions were perfect and included dark, clear and moonless skies with no form of light pollution whatsoever. However, the ZHR should not to be taken as gospel, and you should not expect to actually observe the quantities stated, although 'outbursts' can occur with significant activity being seen.

The observer can also make notes on the various colours of the meteors seen. This will give you an indication of their composition; for example, red

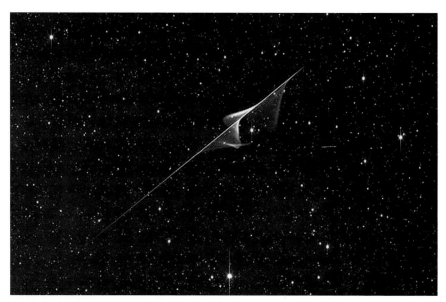

Sporadic meteor captured by the ATLAS (The Asteroid Terrestrial-Impact Last Alert System) Sky survey team on 27 September 2017 using a Schmidt camera, 0.5m, f/2 telescope. (Henry Weiland)

is nitrogen/oxygen, yellow is iron, orange is sodium, purple is calcium and turquoise is magnesium. Also, to avoid confusion with sporadic meteors which are not related to the shower, trace the path back of the meteor and if it aligns with the radiant you can be sure you have seen a genuine member of the particular shower.

Meteor Showers in 2020

SHOWER	DATE	ZHR	PARENT	PEAK	CONSTELLATION
Quadrantids	1 Jan to 5 Jan	120	2003 EH1 (asteroid)	3/4 Jan	Boötes
Lyrids	16 Apr to 25 Apr	18	C/1861 G1 Thatcher	22/23 Apr	Lyra
Eta Aquarids	19 Apr to 28 May	30	1P/Halley	6/7 May	Aquarius
Delta Aquarids	12 Jul to 23 Aug	20	96P/Machholz	28/29 Jul	Aquarius
Perseids	17 Jul to 24 Aug	80	109P/Swift-Tuttle	12/13 Aug	Perseus
Draconids	6 Oct to 10 Oct	10	21P Giacobini-Zinner	7/8 Oct	Draco
Southern Taurids	10 Sep to 20 Nov	5	2P/Encke	10 Oct	Taurus
Orionids	16 Oct to 27 Oct	25	1P/Halley	21/22 Oct	Orion
Northern Taurids	10 Sep to 20 Nov	5	2P/Encke	12 Nov	Taurus
Leonids	15 Nov to 20 Nov	Varies	55P/Tempel/Tuttle	17/18 Nov	Leo
Geminids	7 Dec to 17 Dec	75+	3200 Phaethon (asteroid)	13/14 Dec	Gemini
Ursids	17 Dec to 26 Dec	10	8P/Tuttle	22 Dec	Ursa Minor

Quadrantids

The Quadrantids is a shower that rivals the August Perseids, and peak rates can exceed 100 meteors per hour. The radiant lies a little to the east of the star Alkaid in Ursa Major and the meteors are fast moving, reaching speeds of 40 km/s. A drawback to this shower is that the period of maximum occurs over a very short time period of just 2 or 3 hours. The parent object has been identified as the near-Earth object of the Amor group of asteroids 2003 EH_1 which is likely to be an extinct comet. The moon will be at first quarter this year, resulting in reasonably dark skies which should not interfere with what could be a good show.

Lyrids

These are fast moving meteors with speeds approaching 50 km/s. The rates vary, but typical values of 10 to 15 per hour are recorded. The peak falls on the night of 22/23 April with the radiant lying near the prominent star Vega in the constellation Lyra. The parent of the shower is the long-period comet C/1861 G1 Thatcher, which last came to perihelion on 3 June 1861. The period of this comet is 415 years and it will next approach perihelion in 2280. This year, a nearly new moon should leave dark skies for what may be a reasonably good display.

Eta Aquarids

This is one of the two showers associated with 1P/Halley and is active for a full month between 19 April and 21 May with meteors attaining speeds of around 65 km/s. The radiant lies just to the east of Alpha (α) Aquarii (Sadalmelik) and up to 30 meteors per hour are normally expected during the period of maximum activity, which occurs in the pre-dawn skies of 7 May. However, a nearly full moon will block out all but the brightest meteors.

Delta Aquarids

Linked to the short-period sungrazing comet 96P/Machholz, the Delta Aquarids coincide with the much more prominent Perseids, although Delta Aquarid meteors are generally much dimmer than those associated with the Perseids, making identification much easier for the observer.

The radiant lies to the south of the Square of Pegasus, close to the star Skat in Aquarius and a little to the north of the bright star Fomalhaut in Piscis

Austrinus. Although never very high above the horizon as seen from mid-northern latitudes, the radiant is well placed for those observers situated in the southern hemisphere. The shower peaks during the early morning of 29 July, and in 2020 a first quarter moon may drown out many of the fainter meteors emanating from this shower.

Perseids

The Perseids are fast moving, clocking in at speeds approaching 60 km/s, due to the parent comet (109P/Swift-Tuttle) being in a retrograde orbit. This is a potentially gorgeous spectacle, with meteors appearing as soon as night falls and with up to 80 meteors per hour at their peak, which occurs on the night of 12/13 August. Large fireballs are often observed, with some even seen to cast shadows. A note of caution though – this is a busy time of year for showers, so inexperienced observers should ensure they follow the trajectories of any meteors seen back to the radiant point in the northern reaches of Perseus. Peak activity this year occurs at around the time of the

Perseid meteors captured on 5 August 2016 from Veszprem, Hungary by Monika Landy-Gyebnar. (Monika Landy-Gyebnar)

last quarter moon, which may tend to block out some of the fainter meteors, although the Perseids are bright and active enough to ensure that there should still be a good display.

Southern Taurids/Northern Taurids

This shower is associated with the periodic comet 2P/Encke, with both southern and northern components, the southern hemisphere encountering it first followed by the northern hemisphere later on. The expansive size of this stream is due to a much larger object having disintegrated in the past, leaving a large fragment (2P/Encke) as the main survivor and the remaining debris spread out along the orbit. Larger pieces are believed to lurk within, and many fireballs are seen with this shower. The Taurid radiant lies near the Pleiades (M45) open star cluster. Although there are low ZHRs for both components (between 5 and 10 meteors per hour), they can be beautiful to watch as they appear to glide effortlessly across the sky at speeds of 28 km/s.

Draconids

Also known as the Giacobinids, the Draconid meteor shower emanates from the debris left behind by periodic comet 21P/Giacobini-Zinner. The duration is from 6 October to 10 October, the shower peaking on the night of 7/8 October. The ZHR of this shower can vary. For example, although the hourly rates for 1915 and 1926 appear to have been relatively low, significant displays were seen in 1933 (with hourly rates of several hundred) and 1946 (with thousands of meteors per hour being recorded). Increased activity was also noted in 1998, 2005 and 2012. Radiating from a point near the "head" of Draco, the meteors from this shower travel at a relatively modest 20 km/s. They are generally quite faint, although a waning moon approaching last quarter should leave dark early evening skies on the night of peak activity for what may be a reasonably good display.

Orionids

This is the second of the meteor showers associated with 1P/Halley, occurring between 16 and 27 October with a peak on the night of 21/22 October. The radiant of this shower is situated a little way to the north of the conspicuous red super giant star Betelgeuse in the shoulder of Orion the Hunter, and the

A bright Orionid meteor captured on 18 October 2015 by Andrew Wall from Adelaide, South Australia. The object to lower right of image is a TV aerial. (Andrew Wall)

Orionids are best viewed in the early hours when the constellation is well placed. The velocity of the meteors entering the atmosphere is a speedy 67 km/s. A waxing crescent moon will have set before midnight, which should leaving dark skies for what is hoped to be a good display.

Leonids

This is a fast moving shower with atmosphere impact speeds of 72 km/s and with particles varying greatly in size, including a lot of larger sized pieces of debris with diameters in the order of 10mm and masses of around half a gram. These can create lovely bright meteors that occasionally attain magnitude –1.5 or better. It is interesting to note that each year around 15 tonnes of material is deposited over our planet from the Leonid stream. The shower peaks on the night of 17/18 November, at which time the crescent moon will have set early on in the evening leaving dark skies for what should be an excellent display.

The parent of the Leonid shower is the periodic comet 55P/Tempel Tuttle which orbits the Sun every 33 years and which was last at perihelion in 1998 and is due to return in late-May 2031. The radiant is located a few degrees to the north of the bright star Regulus in Leo. The Zenith Hourly Rates vary due to the Earth encountering material from different perihelion passages of the parent comet. For example, the storm of 1833 was due to the 1800 passage, the 1733 passage was responsible for the 1866 storm and the 1966 storm resulted from the 1899 passage (for additional information see Courtney Seligman's article *Cometary Comedy and Chaos* elsewhere in this volume).

Geminids

The Geminids were first recorded as recently as 1862 which indicates that the stream has only recently been perturbed near to the Earth by Jupiter. The parent of the shower is the object 3200 Phaethon, an asteroid that is in many ways behaving like a comet. Discovered in October 1983, this rocky 5-kilometre wide object is classed as an Apollo asteroid and has an unusual orbit that takes it closer to the Sun than any other named asteroid. Classified as a potentially hazardous asteroid (PHA), 3200 Phaethon made a relative close approach to Earth on 10 December 2017, when it came to within 0.069 AU (10.3 million kilometres / 6.4 million miles) of our planet.

Geminid meteors travel at relatively modest speeds of 35 km/s and disintegrate at heights of around 40 kilometres above the Earth's surface. The shower radiates from a point close to the bright star Castor in Gemini and peaks on the night of 13/14 December. This is considered by many to be the best shower of the year, and it is interesting to note that the number of observed meteors appears to be increasing each year. As far as the 2020 shower is concerned, the nearly new moon will provide dark skies for what may be an impressive display.

Ursids

Discovered by William F. Denning during the early 20th century, this shower is associated with the Jupiter family comet, 8P/Tuttle. It has been noted that outbursts occur when the comet is at aphelion as some meteoroids get trapped in a 7/6 orbital resonance with Jupiter. The shower radiant lies near Beta (β) Ursae Minoris (Kochab). With relatively slow speeds of 33 km/s, the Ursids

This composite image of 7 frames taken by Andrew Wall from Adelaide, South Australia on 15 December 2014, shows a number of Geminid meteors. (Andrew Wall)

will appear to move gracefully across the sky. The moon will be at first quarter which should leave reasonably dark skies, providing observers with ample chance to view what may be a good show.

Article Section

Astronomy in 2019

Rod Hine

ALMA Insights

The Atacama Large Millimeter/submillimeter Array (ALMA) continues to provide amazing images and intriguing insights into many aspects of astronomy. Comprising 66 antennae located on the Chajnantor Plateau in the Atacama Desert in Chile at an altitude of around 5,000 meters above sea level, the observatory allows observations at wavelengths in the range of 9.6mm to 0.3mm. The thin, dry air minimises losses that would occur at sea-level due to water vapour, oxygen and carbon dioxide. The multiplicity of antennae can be moved around and arranged in various configurations spanning about 15 kilometres. Using interferometric techniques gives a spatial resolution of 10 milliarcseconds (10^{-7} radians), or about 5 times better than the Hubble Space Telescope.

Recent results from several different projects at ALMA have shed light on such diverse topics as the formation of protoplanetary disks, magnetic fields in jets from newly-formed stars and huge streamers and galaxy-scale fountains of gas. In all cases the exceptionally fine resolution has revealed details never seen before and of great value to understanding what is going on in the vicinity of these objects. There is doubtless much more to come from this international collaboration which has been claimed to be the most expensive ground-based observatory at a cost of about US$ 1.4 billion. Whilst expensive it is a fraction the cost of say, the James Webb Space Telescope which may well top out at nearly US$ 10 billion by the time it is launched in 2021.

First CMB, now EBL

Since its serendipitous discovery, the Cosmic Microwave Background Radiation (CMB) has formed a cornerstone of theories of the origin of the universe. It was the physicist and polymath George Gamow who postulated the existence of "fossil radiation" or a kind of afterglow from the Big Bang. A chance observation of excessive noise in an old satellite antenna in 1965 gave Arno

This image, the sharpest ever taken by ALMA, shows the protoplanetary disc surrounding the young star HL Tauri. It reveals substructures within the disc that have never been seen before and even shows the possible positions of planets forming in the dark patches within the system. The resolution is significantly better than that achieved in visible light by the NASA/ESA Hubble Space Telescope. ALMA (ESO/NAOJ/NRAO)

Penzias and Robert Wilson the Nobel Prize in 1978 and prompted the detailed investigation of the CMB by land and space-based telescopes, most recently by Planck. After being released when the Universe was about 380,000 years old, the CMB spread throughout the hot primordial hydrogen and helium, cooling all the while from 3,000 K towards its present temperature of 2.7 K.

There followed a long period before the universe had expanded enough and the gases had cooled enough for the first stars and galaxies to form – the

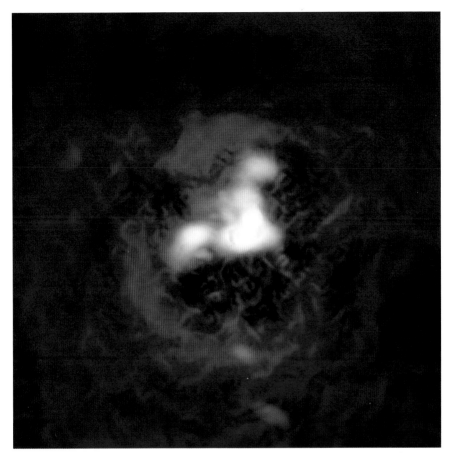

The Abell 2597 galaxy cluster has a supermassive black hole in the central galaxy which is driving flowing fountains of gas. This composite image from three instruments shows cold infalling gas imaged by ALMA in yellow. The red data from the MUSE instrument on ESO's Very Large Telescope shows a spout of hot hydrogen gas erupting from the black hole. The blue-purple is the extended hot, ionized gas as imaged by the Chandra X-ray Observatory. (ALMA (ESO / NAOJ / NRAO), Tremblay et al.; NRAO / AUI / NSF, B. Saxton; NASA / Chandra; ESO / VLT)

re-ionisation at the end of the so-called "Dark Ages" – but soon the light from new stars began to spread out. Once such photons had escaped from their parent stars and galaxies they were destined to travel forever unless involved in an interaction with other matter. Here on Earth our powerful telescopes, peering back billions of years, occasionally detect a few such photons from early stars and galaxies. We see them with large values of red-shift of course but the overwhelming majority of such photons are still travelling at the speed of light, as will most of the photons emitted so far by our own Sun. The term "Extragalactic Background Radiation" (EBL) has been coined to describe these eternal travellers, and recent work has estimated the total number of EBL photons at 4×10^{84}, a staggering number even by astronomical standards! But the CMB photons still outnumber them by a million to one …

So what is the significance of the EBL and what can we learn from it? Firstly, the total energy content forms a significant part of the energy of the whole universe. Secondly, knowledge of the spectral distribution and red-shift of the EBL gives useful information about the earliest era of star formation.

A little thought should reveal the difficulty of studying the EBL – how is it possible to detect and measure this fog of photons at all, aside from the few that arrive in telescopes? It turns out that in the right circumstances, photons can interact with other photons in a process called "pair production". A photon with high energy, typically a gamma ray photon, may have sufficient energy to make a pair of particles such as an electron and a positron. The Einstein relationship $E=mc^2$ tells us that it is possible if the original photon has at least 1.022 MeV energy or wavelength less than 10^{-6} microns. However, considerations of conservation of energy, momentum and quantum parameters forbid a spontaneous creation without the presence of something else. If this photon hits an atom then the electron-positron pair can be formed and the recoil of the atom and the new particles carries away the momentum and surplus energy. However, it is also possible for this interaction to happen with another lower-energy photon, although the chance of it happening is very small. In the terminology, the interaction cross-section is very small, so it takes a lot of photons over a huge distance to give even a small number of interactions – just the conditions of the EBL in the intergalactic void.

This means that gamma ray photons from distant active galaxies or "blazars" can and do interact with a few of the vast number of EBL photons giving an attenuation effect that can be measured. A group known as the Fermi-LAT Collaboration has observed the gamma ray profiles from 739 active galaxies using the Fermi Gamma-ray Space Telescope and from the results has built a model for the evolution of the EBL, and hence the history of star formation. They conclude that star formation reached a peak around 3 billion years after

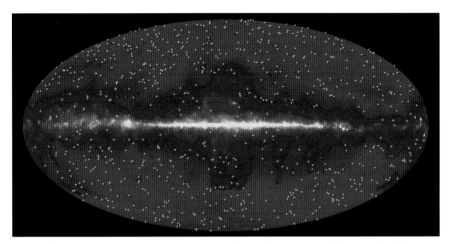

This map of the entire sky shows the location of 739 blazars (green dots) used in the Fermi Gamma-ray Space Telescope's measurement of the extragalactic background light (EBL). The background shows the sky as it appears in gamma rays with energies above 10 billion electron volts, constructed from nine years of observations by Fermi's Large Area Telescope. The plane of our Milky Way galaxy runs along the middle of the plot. (NASA / DOE / Fermi LAT Collaboration)

the Big Bang. According to David Thompson, Fermi's deputy project scientist at NASA's Goddard Space Flight Centre in Greenbelt, Maryland, this provides a useful confirmation of previous measurements. "In astronomy, when two completely independent methods give the same answer that usually means we're doing something right. In this case we're measuring star formation without looking at stars at all but by observing gamma rays that have travelled across the cosmos."

Readers interested by this topic should look at the Royal Society Open Science article at: **https://royalsocietypublishing.org/doi/pdf/10.1098/rsos.150555**

The Event Horizon Telescope and Black Holes

The Event Horizon Telescope (EHT) is not a single instrument but a collaboration of nine radio telescopes, operated together using interferometry to make a world-sized virtual telescope. This long-term project uses dishes spanning from Greenland to the South Pole and from Hawaii to the French Alps operated by eleven institutions with numerous other affiliates. The principal

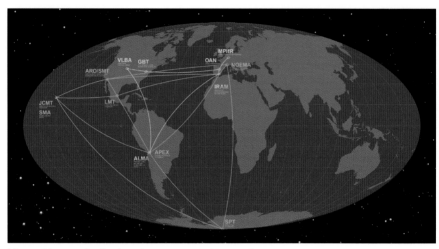

The EHT network of radio telescopes. (ESO / O. Furtak)

aim is to investigate the immediate environment of a black hole with sufficient resolution to discern the actual event horizon. Theory says that this will exhibit interesting effects due to the strong gravity and allow study of general relativity by the dynamic behaviour of matter travelling near light speed.

The EHT has already made measurements of two supermassive black holes, namely Sagittarius A* at the centre of the Milky Way, some 26,000 light years distant, and the much more massive black hole located at the heart of the elliptical galaxy M87. After simultaneous observations by the individual instruments at several different wavelengths, the huge amount of data was recorded on banks of disk drives which are now being combined and processed. Each site has an ultra-precise atomic clock so that the data from all the sites can be synchronised during processing. The production of a detailed image, expected by 2019, will, if successful, be the first direct observational evidence of an event horizon. It is expected to show a dark shadow with a bright crescent of light where the rotation of the black hole drags light along with it. Comparisons between the actual image and the numerous simulations should improve our understanding of the extreme conditions around black holes. According to Heino Falcke of Radboud University, Nijmegen, the Netherlands, one of the collaboration members, it may help to resolve the major difficulty of reconciling the large-scale effects of gravity with the small-scale effects of

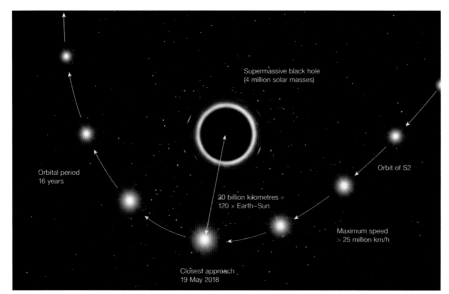

By studying the star S2 passing closely by the supermassive black hole in the centre of the Milky Way, astronomers have successfully tested Einstein's theory of general relativity. This artist's impression shows how the star's light was slightly red-shifted by the very strong gravitational field during the event. (ESO/M. Kornmesser)

quantum mechanics. Somewhere in the image of the event horizon it may be possible to see the point where one or other theory goes wrong.

Sagittarius A* also figures in observations made by another group at ESO, when in May 2018 the star S2 passed very close to the black hole. Shielded by dust from most visual telescopes, the event was observed in the infra-red by instruments on the ESO Very Large Telescope in Chile. One of a group of stars in tight orbits around the black hole, S2 passed at 25 million kilometres per hour through the intense gravitational field and as it did so, its light was stretched and distorted exactly as predicted by Einstein's Theory of General Relativity.

Last Words from Planck?

The Planck Consortium has published the final tranche of papers, known as the legacy data release, from the Planck Space Telescope which was launched in 2009 and operated until 2013. There are no surprises and the current model

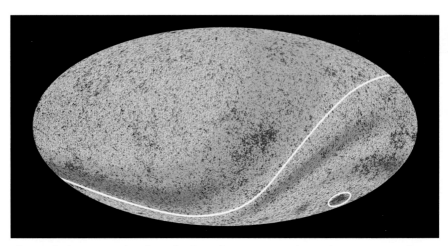

The new high precision data from Planck confirms two anomalies in the CMB previously seen by NASA's Wilkinson Microwave Anisotropy Probe (WMAP). To make them more visible in this image the anomalous regions have been enhanced by red and blue shading. The curved line shows the asymmetry in the average temperatures on opposite hemispheres of the sky, the south being slightly hotter. The circle shows a cold spot that extends over a patch of sky that is much larger than expected. Both features contradict the prediction made by the standard model that the Universe should be broadly similar in any direction we look. (ESA and the Planck Collaboration)

of the universe is confirmed yet again. Only the same few anomalies remain as disclosed in the 2013 results, namely some subtle variations in the signal and the presence of a "cold spot". "This is the most important legacy of Planck," says Jan Tauber, ESA's Planck Project Scientist. "So far the standard model of cosmology has survived all the tests, and Planck has made the measurements that show it."

Whether this really does represent the final part of the story of the origin of our universe and the mystery of Dark Energy and Dark Matter remains to be seen. However convincing the Standard Model seems, the anomalies remain and there may well be some great discovery yet to be made which will reveal some deeper truth – and doubtless pose even more problems for future cosmologists to ponder.

Solar System Exploration in 2019

Peter Rea

This article was written in the winter of 2018/2019. As all the missions mentioned are active, the status of some missions may change after the print deadline. In the case of the InSight mission to Mars, this article was written immediately after landing but before instrument deployment. There has been much activity since the last Yearbook and much of this article takes place in 2018 but continues into 2019.

Gaining an Insight into the Martian Interior

"Touchdown confirmed." The words we all wanted to hear. It came at 7:54pm GMT on 26 November 2018. The next Mars lander had just arrived at Elysium Planitia just north of the last Mars lander, the Mars Science Laboratory. Its journey began on 5 May 2018 from Vandenburg Air Force Base in California, the first time any planetary mission had been launched from the west coast. After

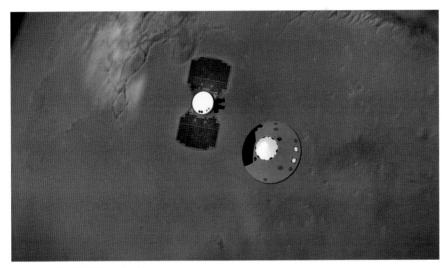

Figure 1. (NASA/JPL-Caltech)

travelling more than 484 million kilometres InSight separated from the cruise stage (Figure 1) that had provided power and navigation on its more than six month journey to the red planet. The landing capsule hit the upper atmosphere at around 19,000 kilometres per hour and 128 km above the surface. Friction with the tenuous atmosphere caused the heatshield temperature to reach 1,500°C and reduced most of the entry velocity. At 11 km above the surface a supersonic parachute was deployed to further reduce velocity. At about 1 km above the surface and travelling at 200 kph the lander separated and fired up its thrusters to slow InSight down to 8 kph for a safe landing. We were able to follow InSight all the way through the atmosphere and onto the surface of Mars due to a remarkable new communications concept. Launched on the same rocket as InSight were two small CubeSats, no bigger than a briefcase. They would travel alongside InSight all the way to Mars. However, instead of aiming for the atmosphere they were targeted to fly past Mars and act as a communications relay. This can be seen in Figure 2. They worked perfectly so all of us back on Earth could follow InSight all the way down to the surface.

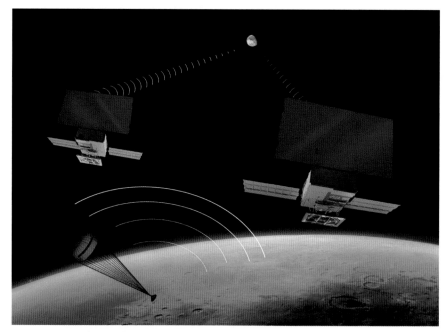

Figure 2. (NASA/JPL-Caltech)

Figure 3. (NASA/JPL-Caltech)

InSight carries two main science instruments. A Seismic Experiment for Interior Structure (SEIS) to listen to Mars quakes and a Heat Flow and Physical Properties Package (HP3) which will send a small drill down into Mars perhaps as much as 5 metres. It was designed to burrow below the Martian surface to measure how efficiently heat flows through the core of Mars. A further experiment called the Rotation and Interior Structure Experiment (RISE) will use the spacecraft radio to provide precise measurements of planetary rotation to better understand the interior of Mars. All previous landers on Mars have studied the surface of the planet. The InSight mission is all about the interior of Mars, an area of science previously unexplored. An artist's impression of InSight with instruments deployed on the surface is shown in Figure 3.

For the record, InSight is an acronym for Interior Exploration using Seismic Investigations, Geodesy and Heat Transport. Don't you just love these NASA acronyms!

Mission website can be found at **http://insight.jpl.nasa.gov**

Our Sun, Up Close and Personal

An ambitious mission to study the Sun left Earth on 12 August 2018. The objective of the mission is to make repeated close passes of the Sun. This is not an easy task. In its annual journey around the Sun our Earth travels with an average orbital velocity of 108,000 kilometres per hour. In order to "fall" toward the Sun a significant part of this orbital velocity must be removed. To accomplish this, the United Launch Alliance Delta 4 Heavy launch vehicle with an addition upper stage was chosen. The energy imparted by this launch vehicle was still not enough to achieve the mission objectives and additional energy in the form of 7 close flybys of Venus is required. The Parker Solar Probe named for astrophysicist Dr Eugene Parker, will travel around the Sun 24 times. In 2025 it will approach to within 6.1 million km of the Sun, the closest any spacecraft has been. The gravitational field of the Sun at that distance is so strong that the probe will fly past the Sun with a velocity of 690,000 kph. The first Venus flyby occurred on 3 October 2018 just 52 days after launch. This flyby reduced the probe's speed relative to the Sun by 10 percent. This equates to about 11,200 kph. This manoeuvre reshaped the orbit so that the probe would fly by the Sun 6.4 million kilometres closer than without the manoeuvre. The first of many perihelia or closest distance to the Sun occurred on 5 November 2018 just 83 days after launch. An artist's impression of the probe passing close to the Sun is shown in Figure 4. Due to the spacecraft's closeness to the Sun and the intense radiation, it was not possible to communicate with it during the perihelion passage. First indications of a successful flyby were received on

Figure 4. (NASA/Johns Hopkins APL/Steve Gribben)

16 November. Parker Solar Probe needed to move well clear of the Sun before a strong signal enabling data return could be picked up. The recorded data was returned to Earth during December. The dates of future perihelia and the dates of the remaining six flybys of Venus can be found on the mission website shown below.

Mission website can be found at **http://parkersolarprobe.jhuapl.edu**

Mission to Mercury

At precisely 02:45 BST on 20 October 2018 an Ariane 5 launcher, flight VA245 sent the BepiColombo spacecraft on a 7-year journey to orbit the innermost planet in the solar system, Mercury. As discussed in the Parker Solar Probe section, sending a spacecraft close to the Sun is not easy due to the intense gravity field of the Sun. In terms of energy requirements, it takes less energy to reach Pluto than is does Mercury. To fly directly to Mercury and go into orbit either requires huge amounts of onboard propellant, which would make the spacecraft too heavy to launch, or we go the long way around and use the gravity of three planets to shape the trajectory. Named for the Italian scientist, Guiseppe (Bepi) Colombo the spacecraft will spend the first year and a half of its journey to Mercury outside of Earth's orbit. On 6 April 2020 the spacecraft will return to Earth for a flyby which will bend the trajectory inward toward Venus for a flyby on 20 October 2020. A second flyby of Venus on 11 August 2021 will send BepiColombo towards its first flyby of Mercury on 2 October 2021. Five additional flybys of Mercury through 2025 will gradually circularise the orbit and reduce the relative velocity between BepiColombo and Mercury, allowing the onboard propulsion system to place the spacecraft into orbit on 5 December 2025.

The spacecraft consists of four elements. Mercury Transfer Module (MTM) used for propulsion and built by ESA, Mercury Planetary Orbiter (MPO) built by ESA, Mercury Magnetospheric Orbiter (MMO) built by JAXA and a sunshade that houses the MMO during cruise. These elements can be seen in Figure 5. The Mercury Transfer Module contains large solar arrays to generate electrical power to operate a suite of ion propulsive thrusters. Whilst generating low thrust, ion thrusters can be operated for long periods of time with high efficiency. On arriving at Mercury, the MPO will be placed into a 2.3 hour orbit and the MMO into a 9.3 hour orbit. A description of the many

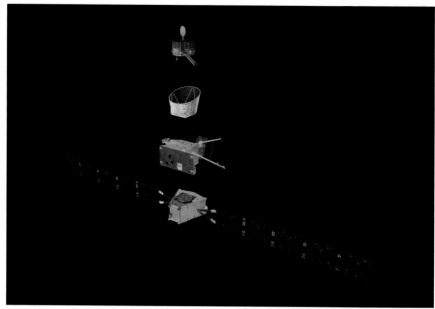

Figure 5. (ESA/ATG MediaLab)

science instruments and science objectives can be found at the mission website listed below.

Mission website can be found at **http://sci.esa.int/bepicolombo**

Asteroid Visit by Japan

On 3 December 2014 the Japanese Hayabusa 2 spacecraft was launched toward the near-Earth asteroid 162173 Ryugu. Like many missions within the solar system the launch vehicle alone cannot impart enough energy to the spacecraft to reach its assigned target. Thankfully there is always free energy to be gained by planetary flybys. Hayabusa 2 returned to Earth on 3 December. This flyby gave Hayabusa 2 the boost in velocity it needed and shaped the orbit so that the spacecraft was able to rendezvous with and go into orbit around Ryugu on 27 June 2018. A close-up image of Ryugu is shown in Figure 6. It is planned that the spacecraft will explore the asteroid with a suite of remote sensing instruments for about a year and a half. The main spacecraft carried with it four small spacecraft attached to its sides. Three of these were Japanese and one was built and supplied by the

Figure 6. (JAXA, University of Tokyo, Kochi University, Rikkyo University, Nagoya University, Chiba Institute of Technology, Meiji University, University of Aizu, AIST)

German Aerospace Centre (DLR). On 21 September 2018 the first two named Rover 1A and Rover 1B were released. Cylindrical in shape and weighing only about 1 kg they both successfully landed on Ryugu. These two were followed by the German MASCOT (Mobile Asteroid Surface Scout), which also successfully landed on Ryugu. Figure 7 shows MASCOT on the surface of the asteroid. Unlike the two Japanese rovers which were powered by solar panels, MASCOT was battery powered and operated on the asteroid surface for about 17 hours. The fourth lander, Rover 2 will be deployed later in the mission.

Figure 7. (JAXA, University of Tokyo, Kochi University, Rikkyo University, Nagoya University, Chiba Institute of Technology, Meiji University, University of Aizu, AIST)

A major mission objective is to collect a sample of the pristine matter comprising the asteroid and return it to Earth. Current planning calls for Hayabusa 2 to leave Ryugu in December 2019 and return to Earth. The sample return capsule is scheduled to parachute to a landing in Australia in December 2020.

Mission website can be found at **http://global.jaxa.jp/projects/sat/ hayabusa2**

Asteroid Visit by the USA

The OSIRIS Rex or Origins, Spectral Interpretation, Resource Identification, Security, Regolith Explorer (wow, what an acronym!) was launched toward the carbonaceous asteroid 101955 Bennu on 8 September 2016. Similar in scope to the Japanese Hayabusa 2 mission, OSIRIS Rex is tasked with returning a sample of Bennu back to the Earth. After utilising gravity assist from the Earth on 22 September 2017 the spacecraft was on a trajectory to intercept Bennu at the end of 2018. During the autumn of 2018 Bennu grew gradually larger in the camera images. These images not only tested the quality of the camera but provided accurate navigational information. Bennu is not a big target. It is only around 500 metres in diameter and rotates once every 4.3 hours. An image of Bennu taken on 16 November 2018 from a range of 136 km is shown in Figure 8. OSIRIS Rex entered orbit around Bennu on 31 December 2018, 2 years,

Figure 8. (NASA/Goddard/University of Arizona)

2 months and 24 days after launch. Operated by NASA's Goddard Space Flight Center, OSIRIS Rex will operate in a close orbit around Bennu for around 500 days. In that time it will map and analyse the asteroid with a suite of remote sensing instruments. At some point during those 500 days the spacecraft will slowly descend to the asteroid surface to collect a sample. This could be anywhere from a few grams to more than a kilogram. Whatever amount is collected will be placed in a sample return capsule. The spacecraft is due to return to Earth and release the sample return capsule on 24 September 2023 where it will land in the Utah Test and Training Range.

Mission website can be found at **http://www.asteroidmission.org**

Juno Continues to Impress

The Juno mission to Jupiter left Earth on 5 August 2011. Juno returned to Earth on 9 October 2013 for a gravity assist that increased its velocity. The spacecraft entered orbit around Jupiter on 5 July 2016. This was a highly elliptical 53 day orbit that took it over the poles with a perijove or closest point to Jupiter of 4,200 km and an apojove of 8.1 million km. By the end of 2018, Juno had completed 17 close flybys of Jupiter after the initial orbital insertion. It had been intended to reduce the 53 day orbit to a 14 day orbit. However, issues with the propulsion system meant that this period reduction manoeuvre could not take place. This will not impact in science return as the mission has been granted an extension. Science return will just take a little longer. End of mission is currently scheduled for July 2021 when it will be deliberately crashed into the atmosphere of Jupiter to avoid any future collision with any of Jupiter's moons.

Readers are encouraged to visit the mission website listed below to view the many magnificent images returned by Juno. One such image showing the moon Io near the limb of Jupiter at the 2 o'clock position is shown in Figure 9.

Mission website can be found at **www.missionjuno.swri.edu**

Ancient Rock and a Distant Traveller

In the cold depths of space, way beyond Pluto and Charon, an irregular shaped lump of rock left over from the creation of the Solar System slowly follows its 296 year journey around the Sun. Since the origin of the Solar System this lonely lump of rock, about 30 km in diameter, will have travelled around the Sun over 15 million times. For the first time in its long existence it will soon be

Figure 9. (NASA/JPL-Caltech/SwRI/MSSS/Gerald Eichstädt/Justin Cowart)

receiving a visitor from a planet in the inner Solar System called Earth. It was first seen by astronomers on 26 June 2014 using the Hubble Space Telescope and given the designation 2014 MU69. Eight years earlier a 478 kg spacecraft called New Horizons was launched from Earth toward an encounter with the Pluto / Charon system. This flyby successfully occurred on 14 July 2015. With all the data returned to Earth, mission managers were looking for another challenge for the New Horizons spacecraft and 2014 MU69 was along the projected flight path. This ancient rock was selected for New Horizons next target on 28 August 2015. A trajectory correction manoeuvre was performed on 22 October 2015 and again on 4 November 2015 placing the spacecraft on a path that would pass 2014 MU69 within 3,500 km on 1 January 2019. During 2018 an observation campaign was conducted as calculations of the orbit of this Kuiper Belt Object determined it would occult a star. Observations made at the time of occultation determined that 2014 MU69 now unofficially named Ultima Thule may be a binary object. The trajectory of New Horizons was fined-tuned during the autumn of 2018 to ensure an accurate flyby. An observation

Figure 10. (NASA/Johns Hopkins University Applied Physics Laboratory/ Southwest Research Institute)

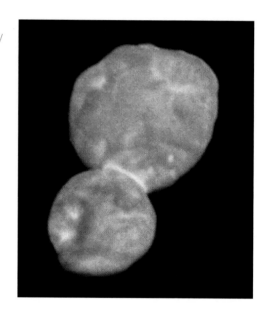

campaign conducted by the spacecraft's Long-Range Reconnaissance Imager (LORRI) looked for any hazards like smaller moons before the final flyby distance was determined. Due to the great distance of Ultima (over 6.5 billion km) it will take all of 2019 to return the data. Images returned shortly after the flyby confirmed that Ultima Thule was in fact a contact binary. At some remote time in the past, two objects came together at relatively slow speed and became attached to each other, as shown in Figure 10.

Mission website can be found at **http://pluto.jhuapl.edu**

Sunset at Dawn

All good things must come to an end. This applies to the NASA mission DAWN currently in orbit around dwarf planet 1 Ceres. Onboard propellant is a finite resource, and this ran out for DAWN in the autumn of 2018. Launched on 27 September 2007 it first entered orbit around asteroid 4 Vesta on 16 July 2011 and departed for Ceres on 5 September 2012. DAWN arrived at Ceres on 6 March 2015 where it has been ever since. With attitude control propellant now exhausted it was officially retired on 1 November 2018. The Sun (metaphorically) has set on DAWN.

Mission website can be found at **https://dawn.jpl.nasa.gov**

Mars Exploration

Mars continues to dominate Solar System exploration. At the time of writing there are nine active missions at Mars, comprising six from the USA, two from Europe and one from India. A NASA website on their Martian exploration can be found at **https://mars.nasa.gov** Further information on the European Mars missions can be found at **http://exploration.esa.int/mars** Details on the Indian Mars mission can be found at **https://www.isro.gov.in/pslv-c25-mars-orbiter-mission**

It would be remiss not to give special mention to two Mars missions from the USA. The Mars Odyssey spacecraft was launched on 7 April 2001 and has been in Mars orbit since 24 October 2001. That's over 18 years at the time of writing. The Mars Exploration Rover, Opportunity, landed on Mars on 25 January 2004. Having landed inside a small crater showing exposed bedrock it explored its local area for a few months before heading off to Endurance crater for six months. It spent most of 2006 to 2008 exploring Victoria crater and then set off on a long three year traverse to Endeavour crater. It has been on the rim of this crater since 2011. In February of 2018 it completed 5,000 Sols (Martian days) on Mars. During the summer of 2018 a large dust storm blanketed most of Mars reducing available sunlight that could be converted by the rover into electricity

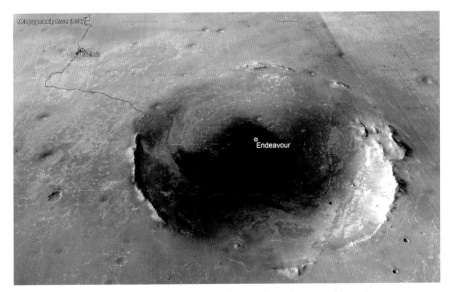

Figure 11. (Google Earth/ESA/DLR/FU Berlin/(G Neukum)/Stuart Atkinson)

using solar panels. With such reduced power Opportunity has not been able to contact Earth since 10 June 2018 the 5,111 Sol on Mars. At the time of writing (late 2018) no signal has been received. Controllers in mission control continue to listen on a regular basis as the dust storm recedes and sunlight on the dusty solar panels increases. Total distance travelled from landing to present location on the west rim of Endeavour crater is 45 km or just over 28 miles. The journey of Opportunity from its landing in Eagle crater to current position on the western rim of Endeavour crater is shown in Figure 11. Remarkable!

Solar System Exploration Programs

NASA's exploration of the solar system has two programs. The Discovery Program which is a series of relatively low cost highly focused space missions. Within this article the InSight mission to Mars is part of the Discovery Program. The program website can be found here: **https://planetarymissions.nasa. gov/missions/discovery**

NASA also has a New Frontiers Program of more ambitious, medium class missions which fall below flagship missions like Cassini at Saturn. Within this article the Juno, New Horizons and OSIRIS Rex missions are part of the New Frontiers Program. The program website can be found here: **https:// planetarymissions.nasa.gov/missions/new%20frontiers**

A NASA website dedicated to planetary missions can be found here: **https:// planetarymissions.nasa.gov**

As always, Solar System exploration continues to excite and inspire and next year promises to be no different.

Anniversaries in 2020

Neil Haggath

John Flamsteed, First Astronomer Royal (1646–1719)

January sees the 300th anniversary of the death of Rev. John Flamsteed FRS (1646–1719), the first Astronomer Royal. The date of his death was in fact 31 December 1719, but this was "Old Style", before Britain adopted the Gregorian Calendar. It equates to 11 January 1720 Gregorian, so the anniversary does indeed fall in this year.

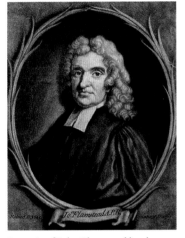

Portrait of John Flamsteed by the German-born painter Sir Godfrey Kneller (1646–1723). (Godfrey Kneller / Wikimedia Commons)

Flamsteed was born in Denby, Derbyshire, but grew up and went to school in the city of Derby. He taught himself mathematics and astronomy by associating with local gentlemen who had such interests. By profession, he became a clergyman, being ordained a deacon of the Church of England in 1674, although his astronomical knowledge and meticulous observing skill were noted in high places.

In 1674, King Charles II appointed a Royal Commission to study methods of determining longitude at sea. Flamsteed was appointed as an assistant to the Commission, to make observations to test the proposed "lunar distances" method. The following year, the King was persuaded by the recently founded Royal Society to establish an observatory for the purpose, and Flamsteed was appointed as "The King's Astronomical Observator", tasked with producing a more accurate star catalogue than any before, for purposes of navigation.

The following year, the Royal Observatory was established at Greenwich; Flamsteed was appointed as its director, and his title became Astronomer Royal. Amazingly, he had to buy his own telescopes and instruments! In 1676, he was admitted as a Fellow of the Royal Society. At the same time, he continued his work as a clergyman; in 1684, he was appointed Rector of Burstow in Surrey. He held that post, as well as that of Astronomer Royal, until his death.

Flamsteed spent an incredible forty years compiling his star catalogue; he refused to release any data until he was entirely satisfied with it. In 1712, however, the Royal Society lost patience. Edmond Halley, who was then Flamsteed's assistant, obtained Flamsteed's data, and he and Sir Isaac Newton published the catalogue without Flamsteed's consent. A bitter dispute ensued between Flamsteed (who was a rather irritable character, due to long-term ill health) and Newton and Halley. Flamsteed obtained 300 of the 400 printed copies of the "pirated" catalogue, and publicly burned them!

In 1725, Flamsteed's widow Margaret posthumously published his own catalogue of 2,935 stars, *Historia Coelestis Britannica*. Ironically, it was never widely used for its intended purpose, as the lunar distances method of finding longitude was rendered obsolete with the development of marine chronometers.

Flamsteed was succeeded as Astronomer Royal by his former assistant (who had become his bitter enemy) Edmond Halley.

The Royal Astronomical Society

This year sees the bicentenary of the founding of the Royal Astronomical Society on 10 March 1820. The anniversary is commemorated elsewhere in this volume, in the article *200 Years of the Royal Astronomical Society* by Sian Prosser.

Eugène Michel Antoniadi (1870–1944)

March sees the 150th anniversary of the birth of one of the greatest of all planetary observers, Eugène Michel Antoniadi (1870–1944). He was Greek, though he was actually born in Turkey. He was born Eugenios Mikhail Antoniadi, in Constantinople (now Istanbul) on 1 March 1870. However, he spent most of his life in France, became a French citizen, and changed his given names to Eugène Michel.

He was first invited to France in 1893 by the great astronomy populariser, Camille Flammarion (1842–1925), becoming an assistant in the latter's private observatory in Juvisy-sur-Orge. In 1909, Henri

Eugène Michel Antoniadi. (Wikimedia Commons)

Alexandre Deslandres (1853–1948) invited him to Meudon Observatory near Paris, to observe with its great 33-inch refractor.

Antoniadi had many British connections; he was a founder member of the British Astronomical Association in 1890, and Director of its Mars Section from 1896, and became a Fellow of the Royal Astronomical Society in 1899.

He was a highly respected observer of Mars; he concluded that Percival Lowell's "canals" did not exist, and in 1929, established a system of naming for Martian features – expanding on the earlier work of Italian astronomer Giovanni Schiaparelli (1835–1910) – which is still in use in modified form today. He also drew the first map of Mercury, although it was flawed due to his mistaken belief that the planet had synchronous rotation, with its rotation period equal to its orbital period. In 1958, when the International Astronomical Union established a formal naming convention for Martian surface features, it adopted 128 names from Antoniadi's 1929 map.

He wrote two famous books in French, these being *La Planète Mars* (1930) and *La Planète Mercure* (1934). Strangely, they were never translated into English until Sir Patrick Moore did so over 30 years later, by which time they were of historical interest only. Of course, his name is best known to amateur astronomers for having devised the Antoniadi Scale of "seeing", a system which we still use today.

Eugène Michel Antoniadi is one of only two astronomers – the other being Schiaparelli – to be honoured with features on three Solar System bodies. There are craters named for him on both the Moon and Mars, and a ridge, Antoniadi Dorsum, on Mercury.

The Curtis-Shapley Debate

A century ago this April, there occurred perhaps the most famous debate in the history of astronomy, the Curtis-Shapley Debate, also known simply as the Great Debate.

Around 1920, one of the most important questions in astronomy concerned what were then called "spiral nebulae"; were they separate galaxies like the Milky Way, at vast distances, or far smaller structures within our own Galaxy? The answer, of course, would have huge implications for the scale and structure of the Universe.

Astronomers had long known of the existence of our own Galaxy, ever since Sir William Herschel deduced its shape over a century earlier. But was it the only galaxy, or was it merely one of many?

On 26 April 1920, two prominent American astronomers held a public debate on the matter at the Smithsonian Museum of Natural History in Washington DC. Harlow Shapley (1885–1972), of Mount Wilson Observatory, believed that our Galaxy was effectively the entire Universe, and that the spiral nebulae were gas clouds within it. A few years earlier, he had estimated the size of the Galaxy, coming pretty close to the value we accept today; he had measured the distances of globular clusters by means of the period-luminosity relationship of Cepheid variable stars, discovered by Henrietta Swan Leavitt (1868–1921) in 1912. He had also deduced the Sun's position within it, from the lop-sided distribution of globular clusters in the sky.

Heber Doust Curtis (1872–1942), of Lick Observatory, held the opposite view – that the spiral nebulae were independent galaxies, of comparable size to our own, and therefore at great distances. However, he also believed our own Galaxy to be considerably smaller than Shapley's estimate.

Each presented an independent paper, presenting evidence to support his view, and then they held a joint discussion in the evening. The following year, they each published a paper in the *Bulletin of the National Research Council*, in which each presented counter-arguments to the other's position in the debate.

Part of Shapley's evidence was that a nova observed in M31, which we now call the Andromeda Galaxy, had outshone the entire galaxy; he thought this was impossible if the "nebula" was a galaxy, as Curtis and others claimed. We now know, of course, that supernovae, which were not known at the time, can indeed outshine their entire host galaxies, and can be seen at intergalactic distances.

He also cited the claim of Adriaan van Maanen that he had observed M101, now known as the Pinwheel Galaxy, rotating on a timescale of a few years, which inferred that the "nebula" could be no bigger than a few light years across. Van Maanen was of course completely mistaken; we now know that galaxies rotate on timescales of millions of years!

Curtis supported his argument with the fact that more novae had been observed in M31 than in the Milky Way – which would be extremely unlikely if M31 was a small part of the Milky Way. He also noted that the dark lanes seen in many galaxies were similar to the dust clouds in the Milky Way.

The Andromeda Galaxy (M31), which featured in both sides of the debate. (Adam Evans/ Wikimedia Commons)

Three years after the Great Debate, the matter was settled once and for all, when Edwin Powell Hubble (1889–1953) determined the distance of M31 as 1.5 million light years (the modern value is actually 2.2 million), thereby proving beyond doubt that it was an independent galaxy. Ironically, Hubble measured the distance of M31 using Cepheids – the same method which Shapley had used to measure the size of our own Galaxy. Shapley graciously accepted that he had been wrong, and continued to do valuable research.

Apollo 13

This year sees the 50th anniversaries of three significant events in spaceflight, the first and best known being the ill-fated flight of Apollo 13.

Launched on 11 April 1970, Apollo 13 was to be the third Moon landing mission. Its Commander, 42-year-old Captain James A. Lovell USN, was making his fourth spaceflight, having previously flown on Gemini 7 and 12 and Apollo 8. Due to the latter, he became the first man to travel to the Moon twice.

The crew of the Apollo 13 lunar landing mission from left to right are: James A. Lovell (Commander), John L. Swigert (Command Module pilot) and Fred W. Haise (Lunar Module pilot). (NASA)

Command Module Pilot John L. "Jack" Swigert, 38, and Lunar Module Pilot Fred W. Haise, 36, were both civilians, and both first-time astronauts.

Swigert was in fact a member of the backup crew; he had replaced Ken Mattingly, who had been dropped from the crew for medical reasons just two days before launch.

Just under 56 hours into the flight, and 328,000 km from Earth, an explosion occurred in one of the liquid oxygen tanks in the spacecraft's Service Module (SM), severely damaging the entire module. When the crew heard "a large bang", Swigert reported to Mission Control with the mother of all understatements – "Houston, we've had a problem here." The cause was later traced to an electrical fault which had gone unnoticed during ground testing.

While the Command Module (CM) was undamaged, it was left without its oxygen and water supplies, power and propulsion. The only power source it still had was its internal batteries, which would be used only for the last few hours of the flight, after the CM separated from the SM. The crew were trapped in a crippled spacecraft.

Thankfully, as they were still on the way to the Moon, they still had their Lunar Module (LM). They powered down the CM, and for the remainder of the flight, used the LM as a "lifeboat", utilizing its oxygen and water supplies to keep themselves alive, while the spacecraft swung around the Moon and headed back towards Earth. They were also able to use the LM's engine to make course corrections.

The LM was designed to sustain two men for up to three days, but it now had to sustain three men for four days. At Mission Control, hundreds of people worked around the clock to work out solutions to the numerous problems, while Ken Mattingly did the same in the spacecraft simulator, working out how to make the spacecraft's limited power supply last long enough to get the crew home.

While the LM had enough oxygen to last, the crew were in danger of suffocation due to the build-up of carbon dioxide. This was removed from the air by chemical filters, which had to be regularly replaced – but the LM did not carry enough filters to last. As the CM and LM had been built by different companies, their filter systems were not compatible; to connect the CM's chemical canisters to the LM's system was, quite literally, a case of putting a square peg into a round hole. A team at Houston worked out how to do exactly that, by making a makeshift adapter from items known to be carried aboard the spacecraft, and the crew were given verbal instructions on how to make it. Ultimately, they owed their lives to a device cobbled from cardboard, plastic bags and sticky tape!

Due to the heroic efforts of hundreds of people on the ground, Apollo 13 made it back to Earth on 17 April, the crew having survived against incredible odds. The mission has rightly been described as NASA's Finest Hour.

At the time of writing (late 2018) Jim Lovell and Fred Haise are both still alive, aged 90 and 84 respectively; I have had the immense honour of meeting both. Jack Swigert, however, died of cancer in 1982, aged only 50.

Two Soviet Space "Firsts"

Also in 1970, the Soviet Union achieved two notable "firsts" in robotic Solar System exploration. The first was to operate the first wheeled vehicle on the Moon, Lunokhod 1, eight months before Apollo 15 carried the first manned Lunar Rover.

Lunokhod 1 was carried to the Moon by Luna 17, which was launched on 10 November 1970, and landed seven days later in Mare Imbrium. Four previous Soviet probes, starting with Luna 9 in 1966, had successfully soft landed on the Moon, and Luna 16 had returned samples to Earth.

The robot rover was designed to operate during the lunar daytime, when it was powered by a solar panel; during each two-week-long lunar night, it was parked and powered down, with a radioisotope power source to keep it warm.

Lunokhod 1 was the first successful rover to explore another world, touring Mare Imbrium for almost 11 months in one of the greatest successes of the Soviet lunar exploration program. (NASA)

Its eight wheels were driven by independent electric motors; steering was done by driving the wheels on opposite sides at different speeds.

Amazingly, while Lunokhod 1 was designed to operate for 90 days, or three lunar days, it actually lasted 322 days! It travelled 10.5 km, returned over 20,000 photographs, and performed soil analysis at several locations.

The rover had to be "driven" remotely from Earth, though it had a sensor-driven automatic braking system to avoid obstacles. Operating it required a team of five men working together, and was so difficult that some of the operators were rumoured to have suffered nervous breakdowns! In spite of such handicaps, the mission was a remarkable success.

Both Luna 17 and Lunokhod 1 itself carried laser reflectors, similar to those left on the Moon by the Apollo crews, which would be used for extremely accurate measurements of the Moon's distance. At the end of its life, before being deliberately shut down, the rover was parked in such a position that its reflectors remained usable.

A second Lunokhod was carried to the Moon by Luna 21 in January 1973.

The second notable achievement was the first landing of a space probe, Venera 7, on the surface of Venus. It was launched on 17 August 1970, and reached Venus on 15 December. Veneras 4–6 had each deployed landing capsules which entered Venus' atmosphere, but were crushed by the immense pressure before reaching the surface. Venera 7's lander was the first to survive intact and soft land on the surface.

The capsule entered the atmosphere at a speed of 11.5 km per second. Friction slowed it to a mere 720 km per hour, then its parachute opened at an altitude of 60 km; it reached the surface 26 minutes later. The lander transmitted data from the surface for only nine minutes, before its telemetry system failed. It measured a surface temperature of 475 °C and a pressure of 90 atmospheres.

Although it returned very little data, Venera 7 had shown that landing on Venus' hellish surface was indeed possible. Nine more Soviet probes successfully landed on the planet during the next 14 years, though none of them survived for longer than a couple of hours.

200 Years of the Royal Astronomical Society

Sian Prosser

The Royal Astronomical Society is a learned society, a scientific community which exists to promote the study and understanding of astronomy and geophysics. In the author's role as custodian of the library and archives of the Society, looking back through the last two centuries a series of key moments stand out in the history of the discipline and the community in which it was involved, from its founding in 1820, to the challenges it faces in the 21st century, via major developments like the birth of astrophysics and the theory of relativity, as well as the changes in membership and the activities the society has run to support the astronomical community. Only a handful of the thousands of astronomers and geophysicists can be mentioned here, but it is hoped that this overview will not only demonstrate the achievements of a few individuals, but

Entrance to the Royal Astronomical Society in Burlington House. (Royal Astronomical Society)

also characterise the contribution that the Society has made to the astronomical community by conferring symbolic credentials in a number of ways, through membership, the granting of awards and prizes, the publishing of research in high impact, prestigious journals, and the organisation of meetings where astronomers can demonstrate their knowledge and research to their peers and to the public.

The late 18th and early 19th centuries saw the increasing specialisation of science into specific disciplines such as natural history and geology, and a growing number of people specialising in these fields who wished to communicate and collaborate with their fellow specialists. The Royal Society was still the pre-eminent scientific organisation, but its membership at the time was limited, and could broadly be characterised as being dominated by aristocratic generalists. This led to the establishment of specialised scientific societies in London, such as the Linnean Society (1788) and the Geological Society (1807) (Basalla et al, 1970). Although observational astronomy was flourishing at this time, John Herschel, commenting decades later on the state of the theoretical underpinning of the discipline in the early 19th century, stated that "Mathematics were at the last gasp and Astronomy nearly so". During this period it was felt that the Royal Society, which once had Isaac Newton as its president, was focused away from mathematics and physical sciences under the long-term presidency of Sir Joseph Banks, a naturalist (Edmunds, 2017a).

It is the Reverend William Pearson (1767–1847) who is credited with proposing, in 1812, a society dedicated to the pursuit of astronomy (Edmunds, 2017c). Eight years later on 12 January 1820, Pearson joined 13 other gentlemen for dinner at the Freemason's Tavern, Lincoln's Inn Fields, where they resolved to establish the "Astronomical Society of London". No professional astronomers from Royal Observatory Greenwich or the universities of Oxford or Cambridge were present at this preliminary meeting; instead the group consisted of men of independent means like Francis Baily, who had made a living as a stockbroker before retiring and dedicating himself to astronomy. The group included members of other professions, such as the clergy (Rev. William Pearson), the law (Daniel Moore) and the military (Captain Thomas Colby). For the next few decades, the membership of the Royal Astronomical Society would continue to be composed of amateur astronomers from the military, clergy and landed gentry, in addition to the increasing number of people who made a living as professional astronomers.

The first president was originally intended to be the Duke of Somerset, but Sir Joseph Banks, who was opposed to the formation of the new society, exerted pressure on the Duke to withdraw. John Herschel's father, Sir William Herschel, rescued the founders from a difficult situation by agreeing to be the first President, on condition that he would never have to chair a meeting as he was at this time advanced in years. The first meeting of both the Council and the new Society took place on 10 March 1820. The Astronomical Society of London became the Royal Astronomical Society after a Royal Charter was signed by William IV on 7 March 1831.

The original objectives of the Society set out by the founders in the "Address" could be said to be a reflection of their interests in "accurate calculation, extensive record-keeping and establishing enduring standards", common concerns to those who work in both astronomy and commerce (Edmunds, 2017d). One of the most ambitious aims, which corresponds neatly to the Society's motto of *quicquid nitet notandum* ("whatever shines should be observed") was to carry out "careful and often repeated observations" in order to discover "the most important secrets in the system of the universe", which "may be concealed under the appearance of very minute single stars"; John Herschel who, as the son of William, knew more than most how arduous such an observation programme would be, was reluctant to include this as an objective, but his peers prevailed. Other aims typical of learned societies were to establish communications with foreign observers, circulate notices of discoveries and forthcoming events, establish a library, propose prize questions, bestow rewards on successful research, and publish research, useful

Presidential portrait of Sir John F.H.W. Herschel. (RAS Papers 113) (Royal Astronomical Society)

observations and tables. The latter two activities, publishing and awards, have been particularly important in the establishing and maintenance of the astronomical community.

From the beginning, the Royal Astronomical Society undertook the publication and dissemination of peer-reviewed papers which were also read during its meetings, building up a credible repository of knowledge which permitted astronomers to understand the current state of research in their discipline, and to advance it. The *Memoirs of the Royal Astronomical Society* was established in 1822, and continued until 1978. *Monthly Notices of the Royal Astronomical Society* (MNRAS) was founded in 1827 in order to circulate timely announcements.

Another of the earliest acts of Council was the decision to have struck a gold medal bearing a portrait of Isaac Newton.[1] The Fellow entrusted with this task was Charles Babbage. He was also the first person to be awarded the Gold Medal by Council for his idea of a "calculating machine" to eliminate errors that plagued the calculating, printing and copying of astronomical and mathematical tables (Edmunds, 2017b). Babbage was the first in a long and illustrious line of recipients of the highest award made by the Royal Astronomical Society. There were interruptions, for example in 1847, described as a "troubled episode" (Dreyer et al, 1923, p. 92). Two astronomers, Urbain Le Verrier at the Paris Observatory, and John Couch Adams at the University of Cambridge, had, unbeknownst to each other, been working on the theory that an undiscovered planet lay beyond the orbit of Uranus; both assumed that the planet would obey Bode's law and both calculated an approximate location for the planet – Adams in 1843, and Le Verrier in 1845. However, although

Gold Medal of the Royal Astronomical Society. (Royal Astronomical Society)

---

1. On the other side of the medal, William Herschel's 40-foot telescope is depicted, but it has not so far been possible to find the record of the decision to feature this design.

Engraving of a portrait of Caroline Herschel by Georg Heinrich Busse in 1847. The astronomer is pointing at the orbit of a comet in a diagram of the Solar System, which includes the orbits of asteroids Ceres, Pallas, Juno and Vesta. (Royal Astronomical Society)

Photograph of the eclipse of the Sun taken at Rivabellosa, Spain in 1860 by Warren De La Rue (reference: RAS MSS Add 146). (Royal Astronomical Society)

Adams had communicated his analysis to the Astronomer Royal, George Biddell Airy, by the time Airy had tasked James Challis, professor of astronomy at Cambridge, with searching for the potential planet, Le Verrier had persuaded astronomers at Berlin Observatory to search for the missing celestial body, which they discovered on 23 September 1846. By 1847, there were so many candidates for the Gold Medal, including Adams and Le Verrier, that Council decided to issue twelve testimonials instead.

Caroline Herschel, the sister of William and aunt of John and a notable discoverer of comets and nebulae as well as the indefatigable assistant of her brother, was the first woman to be awarded the Gold Medal in 1828; no other woman would receive the Gold Medal until Vera Rubin in 1996.

Caroline Herschel and Mary Somerville were both granted honorary membership of the RAS in 1835, but it was not until decades later that women would be able to participate fully in the Society.

From the 1850s onwards, two developments had a major impact on astronomy. One was the increasing use of astrophotography; Warren De La Rue was one of the early expert astrophotographers, whose achievements included

Delegates of the Congrès Astrographique International, 1887. This meeting took place at the Observatory of Paris and resulted in the launch of the Carte du Ciel. (Royal Astronomical Society)

the very successful photograph of a solar eclipse taken during expedition to Rivabellosa, Spain in 1860. It took time for astrophotography to become an established technique in astronomy, and the Society responded by setting up a photographic committee in 1887 which curated a wide range of astronomical photographs for use in publications and to sell to Fellows for teaching and research. During this same year, the Congrès Astrographique took place at the Observatory of Paris, where astronomers from around the world, including Fellows of the Royal Astronomical Society, planned their collaboration in the Carte du Ciel, an international programme to prepare a photographic chart of the heavens using photographs taken by observatories around the globe. This was an ambitious project, similar to the Society's original stated aim to carry out an observational programme of "careful and often repeated observations", but on an even grander scale. Unfortunately not all the participants were able to finish their share of the project, but this collaborative exercise was a forerunner of the International Astronomical Union, founded in 1919, and the goal of the Carte du Ciel would eventually be fulfilled by 21st century astrometry projects like the Hipparcos and Gaia missions.

Another development which had a more immediate effect than photography was the advent of the "New Astronomy", which borrowed from the discipline of chemistry the technique of spectroscopy to analyse the light emitted by celestial bodies, and thus determine their composition. William Huggins (1824–1910), working closely with William Miller, was one of the pioneers in this new field, which became known as astrophysics. Huggins and Miller published their first set of preliminary results in MNRAS in 1863. William Huggins married Margaret Lindsay Murray (1848–1915) in 1875, and they spent the rest of their marriage collaborating on the work of observing and analysing spectra from their observatory in Tulse Hill, London, contributing to an increasingly active field.

Portrait of Sir William Huggins. (Royal Astronomical Society)

Sir William Huggins served for many years on the Council of the Royal Astronomical Society. In addition, both he and Lady Margaret Huggins, who was passionately interested in medieval and renaissance art, influenced the design of a new award, now known as the Jackson-Gwilt Medal, named after Hannah Jackson (née Gwilt), an astronomy enthusiast whose

Portrait of Lady Margaret Huggins. (Royal Astronomical Society)

father and uncle had been Fellows of the Society, and who left a generous gift to the Society on the understanding that it would be used to reward discoveries of new celestial bodies or the invention of instrumentation.

The committee tasked with creating the medal commissioned Ella and Nelia Casella to cast the medal in bronze, with a portrait of William Herschel on the obverse, and the reverse depicting Urania holding an armillary sphere based on woodcuts from early printed books proposed by William and Margaret Huggins. The medal was first awarded in 1897 to the American astronomer Lewis Swift (Vandenbrouck, 2017).

Jackson-Gwilt Medal. (Royal Astronomical Society)

While women played a decisive role in funding, designing and creating the Society's second medal, their involvement in Society affairs was still limited, and, in a reflection of society in general, people from ethnic minorities were even less represented. One of the first Fellows elected to the Society from a country outside Europe and North America was Chintamanny Ragoonatha Chary, head assistant to Norman Robert Pogson at Madras Observatory, India, elected in 1872. Pogson's daughter, Elizabeth Isis Pogson, was proposed for election in 1886, but her name was withdrawn before the balloting stage when it was pointed out that the Society's charter only referred to the male pronoun.[2] A second attempt to elect women was made in 1892, when Annie Scott Dill Russell (later Maunder), Alice Everett and Elizabeth Brown were proposed and balloted, but they were not elected. As a sop, cards of admission were issued to allow women to attend the meetings, which now took place in specially built premises in Burlington House, where the Society had moved in 1874 from Somerset House. The establishment of the British Astronomical Association (BAA) in 1890 arose in part from the need for a forum where women could participate freely in astronomical activity.

The BAA was also founded for amateur astronomers who were discouraged by the cost of RAS membership, and who found the papers too technical. Fortunately the RAS responded more positively the establishment of this

2. Isis Pogson was finally elected in 1920, after the charter and bylaws had been amended.

new association than Sir Joseph Banks had reacted to the founding of the RAS (Dreyer et al, 1923, p. 248). It was in 1916 that women were finally elected to the Royal Astronomical Society as Fellows, rather than honorary members. The first women to be elected included Annie Scott Dill Maunder, and Mary Proctor, a popular astronomer and public lecturer whose lantern slide depicting the spectra of the Sun and various stars is one of the earliest examples of astrophysics outreach. The election of women took place against a background of social pressures from the Suffragette movement and World War I. We shall see that the Great War had an impact on international scientific relations as well.

Portrait of Annie Scott Dill Maunder. (Courtesy of Dorrie Giles)

It is sobering to consult the Society's institutional records and learn who was, in some cases, unjustly passed over for the Society's highest award. The most extreme example of this is Annie Jump Cannon, who was nominated for the medal no fewer than nine times. Albert Einstein was first nominated for the Gold Medal in 1919, but because he was from Germany, a country still in hostile relations with Great Britain just after the war, his nomination was rejected by Council. Embarrassingly, Arthur Eddington had already told Einstein that the award would be given to him. Einstein was finally awarded the Gold Medal in 1926 for his research on gravity and the theory of relativity.

Presidential Portrait of Sir Arthur Eddington. (Royal Astronomical Society)

The war had a negative effect on international communication but pacifist internationalists like Arthur Eddington were able to make connections between astronomers around the world. Arthur Eddington was Secretary during the First World War, meaning that communications from overseas researchers came to him. Because Germany and the UK were at war, no publications were exchanged directly between the two countries. However, Willem De Sitter, an astronomer working in the neutral Netherlands, received Albert Einstein's gravitational theory paper, and communicated these findings to Arthur Eddington. Eddington, realising the significance of the theory when he received De Sitter's communication in 1916, prioritised the publication of this paper in a supplemental issue of MNRAS. In 1919, he was one of the leaders of two expeditions to the island of Príncipe off the west coast of Africa, and Sobral in Brazil, during which the total eclipse was photographed and the deflection of light observed proved the theory of relativity.

In addition to being a leading exponent of the theory of relativity, other major contributions by Arthur Eddington to the discipline include his work on the structure and energy source of stars, subjects on which he clashed repeatedly with James Jeans (1877–1946). In turn, Eddington became embroiled in a famous dispute about the structure of cooling, dying stars with Subrahmanyan Chandrasekhar (1910–1995) in 1935. Many of these debates on fundamental issues are recorded in *The Observatory* magazine, founded in 1877 and which functions in part as the "Hansard" of the Royal Astronomical Society (Shaviv, 2009, Longair, 2006). In 1946, two years after the death of Sir Arthur Eddington, members of Council started considering ways of commemorating his remarkable service to the Society and started raising funds for the Eddington Gold Medal "for outstanding work in theoretical astronomy, more especially in those branches to which Eddington contributed but not excluding celestial mechanics". This was first awarded in 1953.

When the Society was founded, observations were made in the optical wavelength[3] and drawn by hand, later supplemented by photography and spectroscopy. As the 20th century progressed, with improved detectors, larger ground-based telescopes and even telescopes operating in space, astronomers

---

3. Infrared radiation was discovered in 1800 by William Herschel although it took a long time to develop its use for observations.

were able to peer ever deeper into the cosmos and derive useful data about the universe across the electromagnetic spectrum. The Society responded to the exponential increase in research in astronomy and planetary science by greatly increasing the scope of its meetings and the size and contents of its publications. Since 1917 the subject of geophysics, which had been steadily gaining ground in the interests of the Fellows over previous decades, was finally given a dedicated series of meetings in its own right, as well as a *Geophysical Supplement* to MNRAS which is now known as *Geophysical Journal International*. The *Quarterly Journal* was started in 1960 and was published until 1996, providing a home for more general articles, speculative researches, obituaries and historical articles; this was replaced by the new title, *Astronomy & Geophysics*, in 1997, providing a full-colour "glossy" magazine of news and reviews to the Fellowship.

In light of increasing activity by Fellows, the Society has also increased the number of awards available, such as the Annie S. D. Maunder medal for outreach, and Patrick Moore medal for astronomy education, given to commemorate the work that Moore did to popularise astronomy and space science with great lucidity and enthusiasm, using the new medium of television. While the award is relatively recent, Patrick Moore can perhaps the seen as the successor

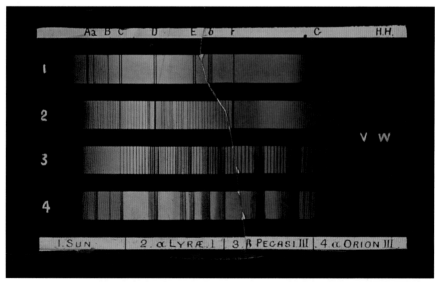

Lantern slide used by Mary Proctor during her public lectures, showing the spectra of the Sun, Vega, Scheat and Betelgeuse. (Royal Astronomical Society)

of popular astronomers of the late 19th century like Mary Proctor, using the new medium of television to explain astronomy and space science to millions of people with lucidity and enthusiasm, as well as being an internationally respected lunar observer (Chapman, 2017). He was an inspiration for so many of our current Fellows, both amateurs and professionals.

While the governance of the Society is dominated by professional astronomers whose working life involves projects like the Cassini-Huygens Mission or the building of instruments to detect gravitational waves, amateur astronomers still have a voice on Council, and just under half of its current members neither work nor study in an academic or professional establishment; some work in the space industry. There are more women represented now, and more BAME members, but the Society is moving from a very low bar and still does not reflect the diversity of 21st century society as a whole.

The Society now performs many additional functions in pursuit of its goals of the encouragement and promotion of astronomy and geophysics, such as funding and facilitating education and outreach, and influencing diversity policy in STEM. A voice for the astronomical community, the Society carries out a range of public policy and advocacy roles, in the past calling for the setting up of a UK Space Agency, and persuading the government to take light pollution more seriously. At the time of writing, as the UK negotiates its future relationship with the European Union, the Society has provided evidence about the potential impact of Brexit on research funding, people and networks, and on immigration policy.

In the conclusion of volume 1 of the history of the RAS, the authors judged that its publications were "the most conspicuous service which the Society has rendered to science" (Dreyer and Turner, 1923). Based on recent surveys, this view prevails among membership today. The income from publishing is used to fund grants, research fellowships and other ways of supporting the community of astronomers and geophysicists. At the same time, the Society is responding to the reasonable requirement for publicly-funded research to be made available for free via open access policies.

Entering the 21st century the Society continues to carry out many of the functions set out in 1820, maintaining a Library, organizing scientific meetings, and publishing journals. So much has been discovered of the "secrets of the universe", but many activities remain constant, because although the Society is an organisation centred on a scientific discipline, it has consistently functioned as a social network for people with a serious interest in the subject, not only providing opportunities to disseminate their results and collaborate with each other, but also to recognise each other's achievements through the granting of awards. As we have seen, sometimes unfavourable decisions have been made, or politics intervenes. Nevertheless, the bicentenary is a perfect opportunity not just to reflect on the last two hundred years in astronomy and geophysics, but to also look ahead positively to the future in the face of uncertainty.

References

Basalla, George, William Coleman, Robert H. Kargon, eds. *Victorian Science: a self-portrait from the presidential addresses of the British Association for the Advancement of Science*, Garden City, New York: Anchor Books, (1970).

Becker, Barbara. *Unravelling Starlight: William and Margaret Huggins and the Rise of the New Astronomy*, Cambridge: Cambridge University Press, (2011).

Chapman, Allan, *The Victorian Amateur Astronomer: Independent Astronomical Research in Britain 1820–1920*, Leominster: Gracewing (2017).

Dreyer, J.L.E and Herbert Hall Turner, eds. *History of the Royal Astronomical Society: 1820–1920*, London: Royal Astronomical Society, (1923).

Edmunds, Michael. 'Founders of the RAS: Olinthus Gilbert Gregory', *Astronomy & Geophysics*, 58:3 (2017), p. 14.

—— Edmunds, Michael. 'Founders of the RAS: Charles Babbage', *Astronomy & Geophysics*, 58:4 (2017), p. 10.

—— Edmunds,Michael. 'Founders of the RAS: William Pearson', *Astronomy & Geophysics*, 58:5, (2017), p. 12.

—— Edmunds, Michael. 'Founders of the RAS: Patrick Kelly', *Astronomy & Geophysics*, 58:6 (2017), p. 9.

Hoskin, Michael (ed). *The Cambridge Illustrated History of Astronomy*, Cambridge: Cambridge University Press, (1997).

Longair, Malcolm S. *The cosmic century: a history of astrophysics and cosmology*, Cambridge: Cambridge University Press, (2006).

Shaviv, Giora. *The life of stars: the controversial inception and emergence of the theory of stellar structure*, Heidelberg: Springer, (2009).

Tayler, Roger (ed). *History of the Royal Astronomical Society: volume 2: 1920–1980*, Oxford: Royal Astronomical Society / Blackwell Scientific, (1987).

Vandenbrouck, Melanie. 'Whatever shines should be observed: astronomical prize medals', *The Medal*, 70, (2017).

The Naming of Stars

David Harper

> The Naming of Stars is a difficult matter,
> It isn't just one of your holiday games;
> You may think at first I'm as mad as a hatter
> When I tell you, a star must have THREE DIFFERENT NAMES.
>
> (With apologies to T.S. Eliot)

The stars have many names. Some are ancient, others more recent. Consider the constellation of Lyra, a familiar feature of northern autumn evening skies. This group of stars was named by the ancient Greeks, who imagined that it resembled the lyre of Orpheus. We know its brightest star as Vega, a name that is derived from the Arabic *wāqi*, which means "falling" or "swooping", in reference to an Arabic name for the constellation, *al-nasr al-wāqi* (the Swooping Eagle).

Most of the brightest stars in the sky have common names which were given to them by Arab astronomers more than a thousand years ago. Six of the seven bright stars that make up the familiar northern winter/southern summer constellation of Orion are known by Arabic names; only one, Bellatrix, has a name from a different language.

Other cultures gave their own names to the stars. In traditional Chinese astronomy, Vega is called Zhi Nü yī which means "First star of the Weaving Girl". Chinese mythology tells that Zhi Nü was a weaver who fell in love with the cowherd Niu Lang, represented by the star Altair in Aquila (the Eagle), but they were separated by the Milky Way and only allowed to meet briefly on one day each year. In Indian astronomy, Vega is called *Abhijit*, a Sanskrit word which means "victorious". The ancient Assyrians knew it as *Dayan-same*, the "Judge of Heaven", whilst the indigenous Boorong people of Australia called it *Neilloan* (the Flying Malleefowl) after a ground-nesting bird that was familiar to them.

Modern astronomers know Vega by other names. It was designated Alpha (α) Lyrae by the German astronomer Johann Bayer (1572–1625) in the early

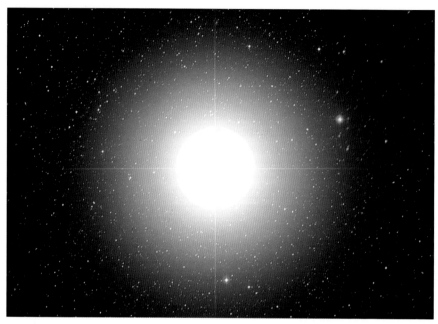

Vega dominates this image, shining brilliantly at magnitude 0.0 and creating prominent diffraction spikes. It was the first star (other than the Sun) to be photographed, and the first to have its spectrum recorded. (Wikimedia Commons/Morigan221)

17th century. Bayer attempted to catalogue the stars in each constellation according to their brightness, using the Greek alphabet. The brightest star was labelled Alpha (α); the second brightest was Beta (β); the third brightest Gamma (γ) and so on, the Greek letter being followed by the genitive form of the constellation name. When he ran out of Greek letters, Bayer used the Roman alphabet.

Vega is also 3 Lyrae, a numerical designation known as the Flamsteed number, although it was actually systematised by the French astronomer Joseph Jérôme Lefrançois de Lalande (1732–1807). Numbers were assigned to the naked-eye stars in each constellation in order of increasing right ascension. As a result, the stars in each constellation crossed the meridian sequentially by their Flamsteed number.

Vega is also listed in numerous all-sky catalogues of stars that have been compiled by astronomers from the late 19th century onwards. These catalogues use numerical designations which follow the Flamsteed convention of assigning numbers in order of increasing right ascension, but without regard to constellation boundaries. A selection is shown in the following table.

CATALOGUE NAME	DATE OF PUBLICATION	DESIGNATION OF VEGA
Bonner Durchmusterung	1859–1903	BD+38 3238
Henry Draper	1918–1924	HD 172167
Yale Bright Star	1930	HR 7001
Gliese Catalogue of Nearby Stars	1957	GJ 721
Smithsonian Astrophysical Observatory	1966	SAO 67174
Hipparcos	1997	HIP 91262
Two-Micron All-Sky Survey (2MASS)	2003	18365633+3847012

Some star names are quite recent. Two curious examples are Peacock (Alpha Pavonis) and Avior (Epsilon Carinae). These previously nameless stars were included in the Air Almanac, a version of the Nautical Almanac which Her Majesty's Nautical Almanac Office (HMNAO) compiled for the Royal Air Force in the late 1930s. The RAF insisted that every star in the almanac must have a name, so names were invented for these two stars by Donald Harry Sadler (1908–1987), the head of HMNAO, and his colleagues. Sadler did not explain how the names were chosen, but the origin of Peacock is an obvious reference to the constellation Pavo. Avior, however, remains a mystery. In Latin, it means "more remote / trackless / untrodden", and this may be a kind of pun, intended to make the reader think of "aviator".

Professional astronomers tend not to use common names of stars when writing articles for scientific journals, except for first-magnitude stars and a small number of unusual objects. The Bayer letter or Flamsteed number is used when referring to naked-eye stars, whilst the Henry Draper catalogue number is preferred for fainter stars down to magnitude 9, and the 2MASS catalogue number is used for stars that are fainter still.

Very few stars are named after people, and these are almost all peculiar objects that are named after their discoverer. The best-known example is Barnard's Star, a very faint low-mass red dwarf which lies about 6 light years from the Sun in the constellation Ophiuchus. It is named after the American astronomer Edward Emerson Barnard (1857–1923), who discovered its extremely high proper motion in 1916. Other astronomers immediately named it "Barnard's proper-motion star". By 1922, it was known simply as "Barnard's Star". However, it was not until 2017, a century after Barnard's discovery, that the International Astronomical Union (IAU) formally approved the name as a recognised designation.

The IAU is the organisation which coordinates and supports international cooperation in astronomy. It has almost ten thousand individual members in more than a hundred countries, and it is recognised by national bodies including the Royal Astronomical Society and the American Astronomical Society. It draws up the rules concerning the naming of all astronomical objects, and these are especially strict when they relate to living persons. The only classes of object that can be named after a living person are comets and minor planets. Comets are named after their discoverer, whilst minor planets may be named by their discoverer to honour another person. Thus, minor planet 2602, which was discovered by American astronomer Edward L.G. Bowell in 1982, is named after Patrick Moore. The IAU does not allow minor planets to be named after political or military leaders until at least 100 years after their death, and not even then in the case of controversial figures. It also does not allow names that might be interpreted as advertisements for commercial products.

Edward Emerson Barnard was an American astronomer who pioneered the use of photography in astronomy. His lasting achievements include cataloguing 370 dark nebulae, and discovering the very high proper motion star which now bears his name. (Wikimedia Commons)

The IAU has recently given its approval to new names for a number of stars which are known to have exoplanets. Its "Name Exo-Worlds" campaign in 2015 invited the public to suggest names for the stars and their planets. Amateur astronomy societies around the globe responded with many suggestions from mythology, history and literature. As a result, there are now stars named after the Polish astronomer Copernicus, the Spanish writer Cervantes, a mythical crocodile king named Chalawan from Thai folklore, and Ogma, the Celtic deity of eloquence, among others.

In 2016, the IAU established the Working Group on Star Names (WGSN) under the aegis of its Division on Education, Outreach and Heritage. The Working

Group is an international panel of experts in stellar astronomy, astronomical history and cultural astronomy. Their remit is to establish guidelines for the naming of stars, to search worldwide astronomical history and culture for star names, and to maintain and publish an official IAU catalogue of star names.

Regrettably, there are commercial companies which offer to name stars after people. For a payment ranging from £15 ($20) to well over £100 ($140), they promise to send an elaborate certificate (suitable for framing!) and to enter the name into an official-sounding "star register". This, they suggest, is an ideal way to mark a birth or to create a memorial in the sky for a deceased loved one. It is sometimes implied that the loved one's name will be kept alive as astronomers use it when referring to the star. This is untrue, of course. Astronomers do not recognise these so-called "star registries" and deplore their activities.

If you know someone who is thinking of paying a so-called "star registry" to name a star after a loved one, please tell them not to waste their money. Nobody will ever use the name to refer to a star. If they are seeking solace after bereavement, you could suggest that they might consider donating the money to one of the many worthwhile charities or to a local community project.

Further Reading

• *Star Names: Their Lore and Meaning* by Richard Hinkley Allen (1838–1908) is the classic reference work on star names. Allen provides a rich historical account of the names of bright stars and the constellations, drawing heavily on Greek and Roman mythology and European literature. The book was first published in 1899 and is still available as a Dover reprint.

• *A Dictionary of Modern Star Names* by Paul Kunitzsch and Tim Smart was published in 2006, and whilst it draws upon more recent scholarship than Allen, it lists only 254 stars and is a much slimmer volume at 66 pages, compared to Allen's 563 pages.

• The International Astronomical Union has published its official position on companies that offer to sell star names: **www.iau.org/public/themes/ buying_star_names**

Astronomical Sketching

Steve Brown

"The universe is full of magical things patiently waiting for our wits to grow sharper."

Eden Phillpotts

Why Astronomical Sketching?

Sketching is a wonderfully rewarding method of recording your astronomical observations and this applies to the novice and experienced astronomer alike. Sketching at the eyepiece has an immediacy often lacking with electronic imaging. One can quickly and easily sketch the view through the eyepiece and with the addition of a few notes, create a permanent record of the observation. Sketching needs no specialist equipment, simply a pencil and a piece of paper. No need has the astronomical sketcher for a whole suite of electronic equipment that requires setting up, calibrating, aligning, protecting from moisture and usually cajoling into working together correctly. Then, if all this does work as it should, the resulting images need stacking, aligning, filtering and processing using specialist software to get the best results. It is little wonder that people starting out in astronomy can be put off by the mere prospect of taking images. Sketching also appeals to the innate creativity that we all possess to some degree. It can be immensely satisfying putting pencil to paper to make a record of what you have actually seen with your own eyes. Producing the sketch makes you process the image at the eyepiece, giving an instant record of what you've seen. How many of us have astrophotography images languishing on a hard drive, without any accompanying records or notes? The experience of capturing these images is easily lost as time moves on and memory fades. With sketching, you are right there at the eyepiece, noting sights and events as they happen and therefore creating an accurate record you can refer back to in years to come. It is perhaps a sad truth of these technological times that many experienced imagers and indeed, many professional astronomers, never see

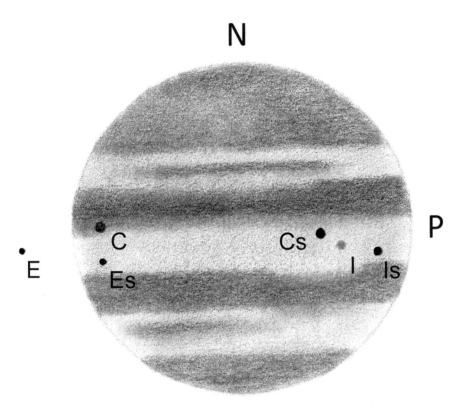

Sketch of a Jupiter triple shadow transit event on 24 January 2015, observed with a 130mm Newtonian reflector at 100x magnification. The sketch shows the Galilean moons Europa (E), Callisto (C) and Io (I) and their shadows - Europa (Es), Callisto (Cs) and Io (Is) - falling on the cloud tops of the planet. (*Drawing by the author*)

the objects they capture with their own naked eyes. Sketching is a wonderful pastime because it gets you 'out there' and really connected to the Universe.

So, what do you need to get started?

Sketching Materials

These are some of the materials and equipment you'll find useful for astronomical sketching:

- A range of pencils of different hardness ratings (at a minimum, you should aim for H, HB, 2B and 4B).

- A good, metal pencil sharpener.
- A kneadable eraser.
- An eraser shield.
- Blending stumps, also known as artists smudge sticks or tortillons.
- A supply of white paper or thin card, preferably acid-free as this does not discolour over time.
- A sturdy clipboard.
- A variable intensity red light torch.

The author's sketching clipboard. This is actually two clipboards glued together to produce an extra-sturdy platform for sketching. The flexible torch holder is a small microphone stand attached to the clipboard. (*Steve Brown*)

- A gooseneck or flexible clamp capable of attaching to the clipboard and holding the torch.
- Masking tape (cover your torch with this as it makes the light from it more diffuse and uniform, making sketching easier).

In addition to the sketch, you should also consider writing some accompanying notes.

Observation Notes

Noting down a few details to accompany your sketch is very useful, as this transforms it from a simple drawing to a permanent astronomical record. You should aim to record the following for each object observed:

- Date.
- Start and finish time in Universal Time (UT).
- Name and type of object.
- Constellation in which the object is located (or alternatively, the position in RA and Dec).
- For planets, the disc diameter and phase (if applicable).
- The location of the observation (with latitude, longitude and altitude if possible).
- Observer(s).
- Instrument used (size, type and mount).
- Eyepiece(s) used and the associated magnification and field of view.
- Seeing conditions at the time (see 'Seeing Conditions' below).
- Sky transparency at the time as a limiting magnitude reading (see 'Limiting Magnitude' below).
- Brief description of the observing session.
- Any filters used and their effect.
- The weather conditions, including wind strength and the temperature if possible.
- For lunar observations, the Moon's phase.
- The size (in arc seconds) of the object and its brightness.

In addition to a description of the observing session, your notes should also record what didn't work. For example, maybe you tried a certain filter to bring

SUBJECT: M37　　　　　　**DATE:** 28/10/2016

W

N

NOTES:

This open cluster was a very different one to M36. It looked roughly heart-shaped to me, a beautiful dusting of stars against the blackness of space. When I used averted vision the cluster brightened considerably, with the brighter stars within the cluster really leaping out as if a switch had been flicked! This is a very beautiful cluster and a stunning sight.

Constellation: Auriga

Magnitude: 5.6

Classification: Open cluster

Observer: Steven Brown

Time: 23:32 — 23:44 UT

Location: Back garden

Instrument: NexStar 6SE (6" SCT)

Eyepiece: 32mm Plössl

Power: 47x　　　Filter: None

Field of view: 1.1°

Conditions: Clear but more high mist noticeable now. No breeze. Temperature 4.6°c.

Seeing: Moderate (3) - Antoniadi scale

Transparency: Good, LM 5.28

An example of a completed observation sheet by the author.

out the belts of Jupiter but it ended up making the view worse. These details can be invaluable in improving your sketching abilities and skills as well as enriching the final record.

There are lots of different ways you can record these details and many amateur astronomers use their own custom-made note templates. I created my own using Microsoft PowerPoint. Each sheet is A5 in size so I can print two per sheet of A4. I print onto 160gsm card to give my notes a bit more durability and store these in A5 presentation folders.

Some excellent templates for recording different astronomical objects can be found on the website of the British Astronomical Association - look under 'Sections'. See 'Resources and Further Reading' below for more places to find templates. Further templates can be found through an online search.

Some astronomers like to draw directly onto their notes at the eyepiece and then produce a final, higher quality drawing on the same page later on. This is a good idea as it results in a 'whole picture' record of the observation. It does have some drawbacks though. Often mistakes can be made in the less than ideal conditions in which astronomy takes place. Cold and damp conditions in the middle of the night are frequently not kind to paper notes. To mitigate this, I use a summary sheet for recording notes at the eyepiece and then transfer these notes to a final, permanent record. This has a further advantage in that I can scribble all kinds of messages and annotations to myself about the observation that can then inform and improve the final sketch.

Seeing Conditions

Astronomical seeing is the blurring and distorting effect of the atmosphere when viewing an object through a telescope. There are many different ways of recording this effect, a common one being the Antoniadi scale, which categorises the seeing conditions under which astronomical observations are carried out.

Recording seeing in this way is useful when looking at your notes as an historical record. Another astronomer will instantly know approximately what you meant by 'poor seeing' or 'perfect seeing', even if their own personal experiences may be slightly different from your own. It is also useful for you to look back at your sketches and see why a particular drawing was good or poor, given certain conditions. As you gain experience you will get to know what to expect through the eyepiece from looking back at your notes.

Limiting Magnitude

The limiting magnitude is a technique used to determine the darkness of the night sky at your location. In addition to this, it is also indicative of the clarity, or transparency, of the night sky. It is simply a figure equating to the faintest star you can see with the naked eye. This is an important figure to note in your records as it tells the reader something about the sky conditions at the time of the sketch. Like seeing, it is a subjective figure but one that should be recorded faithfully for your records to be useful to yourself and others in the future.

There are many different ways of determining the limiting magnitude. One easy method is to use the main stars in the constellation of Ursa Minor. These display a good range of magnitudes, from Polaris at magnitude 2 to Eta (η) Ursae Minoris at magnitude 5. There are also other scales one can use, such as the Bortle Dark-Sky Scale. Personally, I favour the method used by the International Meteor Organisation as part of meteor observing. For details see: **www.imo.net/observations/methods/visual-observation/major/observation/#lm**

All these notes and details are all very good, but they are nothing without the actual sketch…

Basic Sketching Techniques

There are some simple drawing techniques that are frequently used when sketching many different astronomical objects. Regardless of what you are sketching, you should practice the following to help improve your drawings:

Fill

With some objects, such as the Sun, you may want to create a background layer for your sketch. This can represent the surface of an object or a base layer from which to build up other features. You can also erase portions of this layer to revel the white paper beneath, as a way of representing brighter areas of an object. To create the effect, load one of your fingers with graphite from a 4B or greater pencil. The best way of doing this is to heavily shade a scrap of paper and then rub your finger on it. Once done, transfer the graphite to the sketch using circular motions to ensure an even transfer. Start in the centre and work your way to the edges. You should begin with a little graphite and build up the shading to the desired level.

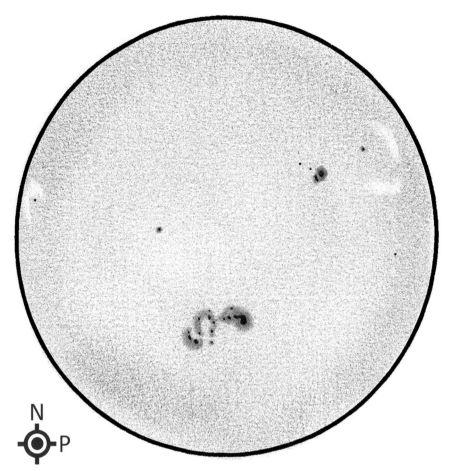

Sketch of the Sun through a white light filter on 7 August 2015 using a 150mm SCT at 60x magnification. A fill technique was used to create the background tone of the solar disc. The texture of the paper has resulted in a good representation of the granulation of the solar 'surface'. (*Drawing by the author*)

Shading

Many objects, such as comets or nebulae, require you to shade a specific area and there are a couple of ways of doing this. You can use a HB or B pencil at a very shallow angle and with light pressure to gently apply a layer of graphite. Then, smooth this with your finger or a blending stump. Apply subsequent layers until your shading matches the object you are observing. Another

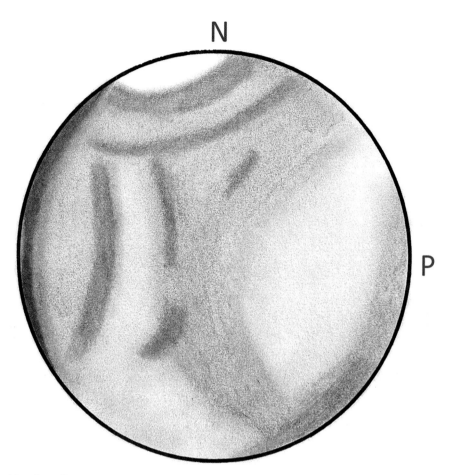

Mars just after opposition on 2 May 2014, observed using a 130mm Newtonian reflector at 163x magnification. Different levels of shading were required to represent the surface features observed. Note the white polar cap. (*Drawing by the author*)

method is to apply graphite to a blending stump or your finger from a piece of scrap paper and then transfer this to your sketch. Typically, a blending stump will allow more precision drawing than a fingertip but covers a smaller area.

Blending

Closely related to shading, blending is simply creating a uniform transition between two or more areas of different colour or tone. It is a useful skill for rendering nebulae and markings on planets. To create the effect, just gradually reduce or increase the amount of graphite applied when creating a shaded area. This can be done by changing the pressure with which you apply the pencil to the paper. This also works when using a blending stump or your finger.

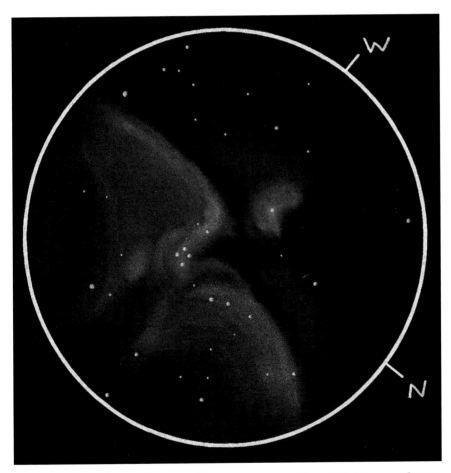

M42, the Orion Nebula on 26 November 2016, observed using a 150mm SCT at 60x magnification. This object requires good use of shading and blending techniques. (*Drawing by the author*)

Stars

To create accurate stars on a sketch you will need a specially sharpened HB pencil. To prepare this, sharpen a pencil with a very good sharpener to a fine point. Then, twist the tip of the pencil on a piece of scrap paper with the aim of producing a uniform, round point. This is so that you can draw stars as circular dots.

Once you are ready to sketch, lightly mark the brighter stars in the field of view. By using light marks you can easily erase any that you have positioned

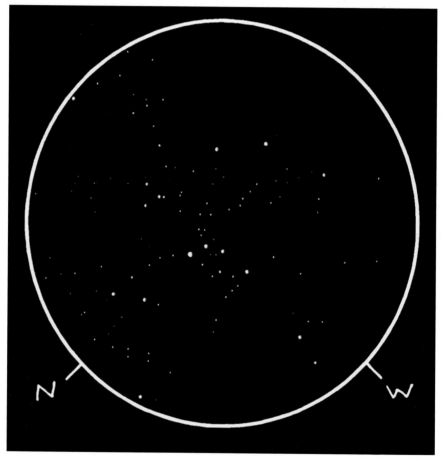

Open cluster M41 in Canis Major, observed on 28 December 2017 using a 150mm SCT at 47x magnification. This sketch shows how different sized dots can be used to represent brighter and fainter stars. The largest dots in this image represent stars around magnitude 7, while the smallest represent stars around magnitude 10. (*Drawing by the author*)

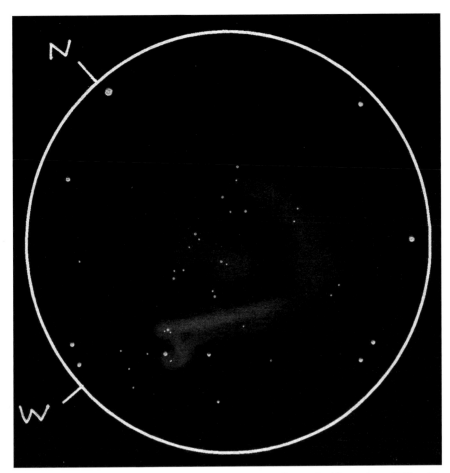

M17, the Omega Nebula in Sagittarius on 5 August 2018, observed with a 150mm SCT at 47x magnification. Many of the techniques described in this article went into producing this sketch. (*Drawing by the author*)

incorrectly. Once you are happy with these, mark any remaining stars. Use the brighter stars as reference points to accurately position the other stars. A useful way of doing this is to imagine various basic shapes formed from different stars and their relation to other stars in the vicinity. Once you are happy with the position of all your stars, use your prepared pencil to gently increase the size and intensity of the dots representing them. Draw larger dots for brighter stars and smaller dots for fainter ones. Once you have all the stars in place you can move on to render any other objects in the field of view, such as nebulae.

When you are sketching at the eyepiece, there are certain techniques you can employ to improve your drawings.

Observing Techniques

Firstly, let your eyes adapt to the dark for at least 20 minutes before starting your sketch. This will allow you to see a lot more. You should also choose a suitable eyepiece for your target object, which should fit comfortably in the field of view. The object doesn't necessarily have to be in the centre of the field of view though. It is fine to position the object with some surrounding stars or other objects of interest. This can provide an important context for the sketch. Also, if too close a view is used then the object can lose context and look less interesting. Of course, this does not always apply as it depends on the object you are sketching. Choose the view that you like the best. It is also important to choose an eyepiece that gives the best magnification for your local conditions. Often the result of increased magnification is just to make the object a blurry mess. Use the eyepiece that gives the best, most stable view for your chosen object at the time.

Before you put pencil to paper let yourself get used to the view for a few minutes. Really take in what you can see. You'll often find that as you relax and observe the object, an increasing amount of subtle detail will be revealed. Try to be aware of how the objects you can see relate to each other. If it is a star field you are looking at, note how the stars are positioned in relation to each other and their respective brightness. If it is a deep sky object such as a nebula, note the varying levels of brightness for different parts of the object. If it is a planet, note the locations and brightness of any surface or cloud features. Only start to sketch once you have taken the time to study the object and really observed it rather than just looked at it. By taking this time at the start, your final sketch will be richer as a result.

Averted Vision

This is a very useful trick to reveal faint detail on any object and is most useful for deep sky observing. Instead of looking through the eyepiece directly at an object, look a little to the side of it. This means the light entering your eye hits a part of your retina that is more sensitive to faint light, which you would normally miss. It is a very powerful technique and under good conditions, objects at the limit of your perception can suddenly appear much brighter and therefore easier to sketch.

Movement Perception

Part of our visual system responds especially well to moving objects and this can be used to aid sketching in a similar way to averted vision. To reveal particularly faint detail on an object, try knocking the telescope very slightly.

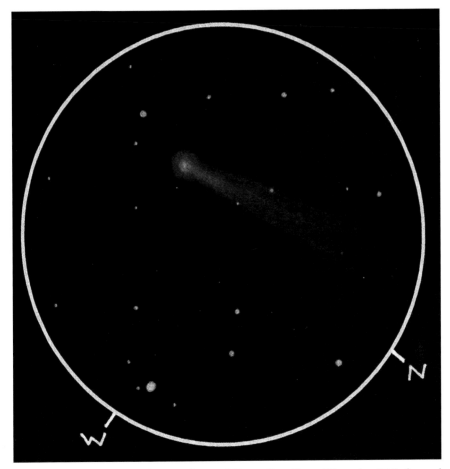

Comet C/2013 R1 Lovejoy in the constellation of Corona Borealis on 6 December 2013, observed with a 130mm Newtonian reflector at 20x magnification. This sketch required intentional movement of the telescope and averted vision to reveal the subtle detail of the comet's tail. (*Drawing by the author*)

This will wobble the object a little in your field of view and creates the illusion that the object you are looking at is moving. This stimulates your motion perception and makes your eyesight more sensitive to light from the 'moving' object. Like averted vision this works best with faint targets such as nebulae and the tails of comets.

Cardinal Directions and Planetary Movement

To finish your sketch, you should mark at least two of the cardinal directions, i.e. north, south, east and west. The easiest way of doing this is to switch off any tracking your telescope is doing and watch which direction the stars drift to. The point at which the stars exit the field of view is west. If you nudge the telescope towards the celestial north, i.e. Polaris, then stars that enter the field of view will do so from the north. For those in the southern hemisphere, nudge the telescope south and the stars will enter from the south. The north (or south) position will be 90 degrees either side of the west point. The exact placement depends on the type of telescope you are using, as they all alter the view in some way. For a telescope with an even number of internal reflections, such as a Newtonian, Cassegrain or reflector without a mirror diagonal, north will be counter clockwise from west. For a telescope with an odd number of reflections, such as a Cassegrain or reflector using a mirror diagonal, north will be clockwise from west.

When drawing the planets, you should also be aware of how your telescope alters the orientation of the image. You should mark at least the north on a planetary image (N) as well as the 'preceding' limb (P). Preceding is the direction in which the planet is rotating, so features toward that edge will move around the disc as time moves on and be lost. Conversely, features will rotate into view from the 'following' edge (F).

Conclusion

Hopefully this article has given you the basics with which to start your own sketching record. Exactly how you go about this is up to you. The notes you read here should be viewed as guidance only, rather than instructions. Astronomical sketching is very much a personal activity, recording your view of the Cosmos in your own way. As you continue to sketch you will find you develop your own personal style, which is as it should be. It is your own view and interpretation of the Universe and you should record your observations in your own way. If you are not satisfied with your results at first, keep practising. By continuing to sketch you will also become a better observer, which is a skill that will serve you well for years to come. There is a whole universe of objects out there, just waiting to be seen by your keen eye and recorded by your artistic skills. Get out there and get sketching!

Resources and Further Reading

Astronomical Sketching - A Step-by-Step Introduction, Richard Handy et al.
An excellent book that provides in-depth guidance on sketching many astronomical objects.

The British Astronomical Association website **https://britastro.org** contains many useful sketching articles and templates.

Jeremy Perez is an accomplished visual observer and sketcher. His website The Belt of Venus at **http://www.perezmedia.net/beltofvenus** contains many useful tutorials and an excellent templates page.

Dark Matter and Galaxies

Julian Onions

Dark matter, right from the beginning, feels like something that has been made up to fix a problem, and in truth it has. It is somewhat of a "fix" that makes a number of observed things right, which we shall go into in the following text. It must be said up front, that the idea of dark matter may well turn out to be wrong in the end. With that said, it should also be understood that the concept is embraced by probably more than 99% of astronomers as accepted and is used to underpin many ideas and models. So, if it does turn out to be wrong, there will be a lot of head scratching going on!

It is also not the first time we have made a leap of faith in physics. In 1930, theoretical physicist Wolfgang Pauli (1900–1958) proposed a new particle called the neutrino to "fix" an issue with radioactive decay. Making up a new particle meant the existing laws of physics would again work correctly in this case, but it was not until 1956 that one was directly detected. There is a similar story to be told about the Higgs boson, predicted in the early 1960s, but not found until 2012. Although this doesn't prove anything, it at least shows we are in good company.

Why do we think there is dark matter?

The initial hints that there was something missing came from Dutch astronomer Jan Oort (1900–1992), after whom the Oort cloud is named. He was working out where the Sun was in relation to the rest of the Milky Way. He eventually showed that we were not at the centre, and were rotating around the centre at maybe half the total distance. However this showed up a problem, in that the Sun was rotating around the Milky Way faster than it should be. With the stars he could see, and the associated gravity, there was not enough mass to hold the Sun gravitationally in its galactic orbit. It was like swinging a ball on a string around your head, but choosing cotton rather than string. He put forward the suggestion that there was either a considerable number of missing stars that he could not detect, or else there was something else there providing the necessary gravity to hold everything in place.

At about the same time, the 1930s, Fritz Zwicky (1898–1974) a Swiss-American astronomer was doing something roughly similar but on a much larger scale. The stories about Zwicky are legendary in astronomical circles, but he was clearly very clever, if not particularly easy to work with. He was looking at the Coma Cluster, a collection of more than 1,000 galaxies lying reasonably close by. He effectively weighed the cluster by two methods, the first being by looking at their movement around the cluster based on the virial theorem (which relates the measured kinetic and potential energy in any system that is not undergoing change, and hence allowing the mass to be worked out). The other method uses the mass to light ratio – how much light is observed which can be related to how many stars there are producing it. The first figure was 400 times the second, which is a large error, even by astronomical standards. Zwicky was the first to coin the term "dark matter" relating to this missing

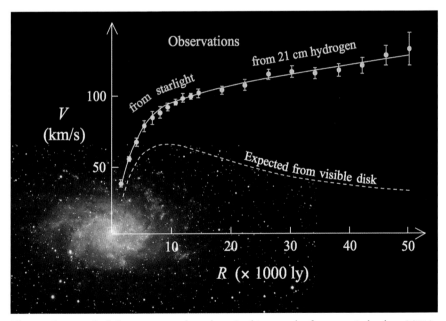

The rotation speed of the stars around a galaxy (in this case, the face-on spiral galaxy M33 in Triangulum) is shown, together with what would be expected from the visible stars in the galaxy. They clearly diverge greatly after the initial agreement. There are two ways to correct this. Add in a lot of extra mass to a large distance from the centre of the galaxy (beyond where the stars end) or change the laws of physics. (Wikimedia Commons/Stefania.deluca)

quantity. Today we have much better estimates of a lot of the figures that went into his calculation, and the factor is much smaller, but still significant.

The next step along the road came with American astronomer Horace Welcome Babcock (1912–2003) in 1939, when he was studying the Andromeda Galaxy (M31). He showed there was an issue in the rotation curve of the galaxy, and that stars in the outer part were again going far too fast for the amount of matter within the galaxy to hold them in. He did not attribute this to dark matter, but instead came up with other possible explanations.

It was not until the 1960s and '70s that Vera Rubin (1928–2016), widely regarded as a pioneer on galaxy rotation rates, looked in more detail at the rotation of M31 with a much improved spectrograph, and showed much more conclusively that the rotation did not follow the expect profile. Furthermore, radio telescopes were now entering the picture and mapping hydrogen clouds that extended outside the visible portion of M31, and they showed the same problem, but to a much greater distance.

Galaxy clusters again came back into the story as new techniques were developed. As well as the virial method used by Zwicky, it is also possible to give a good estimate of the mass of a galaxy from the X-ray emissions coming from it. Using X-rays to measure the gas temperature in a galaxy cluster can again be related to the mass required to keep it in balance. Once again, the figures were out …

Albert Einstein famously worked out that mass causes light to bend, and galaxy clusters being some of the most massive objects in the universe cause

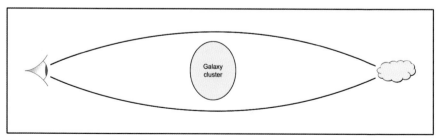

The large mass of stars in a galaxy cluster acts as an imaging device, so that light from a remote source passing nearby bends around it (in much the same way as light is bent in a magnifying glass). This can lead to a distant galaxy being magnified and distorted. Using general relativity, we can work out how much mass is required for the distortion observed. This is always more than astronomers can easily account for by the visible stars and galaxies. (Garfield Blackmore)

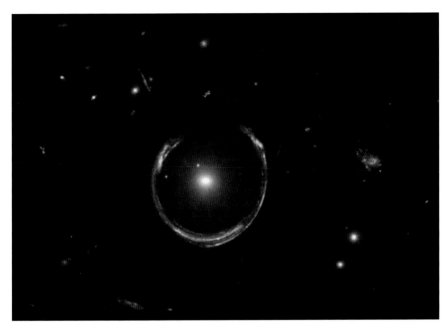

Gravitational lensing is a process which acts like a glass lens, bending and magnifying light from behind. In this image, the light from a distant blue galaxy (appropriately nicknamed "the Cosmic Horseshoe") is bent and stretched into a nearly complete ring (a so-called Einstein Ring) around the foreground red elliptical galaxy LRG 3-757. Amongst other things, this process provides astronomers with a way to measure how much mass there is in the foreground galaxy. (ESA/ Hubble and NASA)

light to bend significantly. This leads to gravitational lensing, where the light from background galaxies is seen to be bent around these clusters, which gives us another way to measure the mass of a cluster. All these methods lead to similar figures that there is more mass there than can be accounted for by the visible matter. In general this is that dark matter outweighs regular matter by about 5:1.

More evidence comes from the cosmic microwave background (CMB) and the various satellites measuring and studying it, such as Cosmic Background Explorer (COBE), the Wilkinson Microwave Anisotropy Probe (WMAP) and the Planck space observatory. Careful analysis of the results from these missions leads to a clear indication that there is more to the universe than we can see. In particular an indication of something that is matter-like, but not behaving

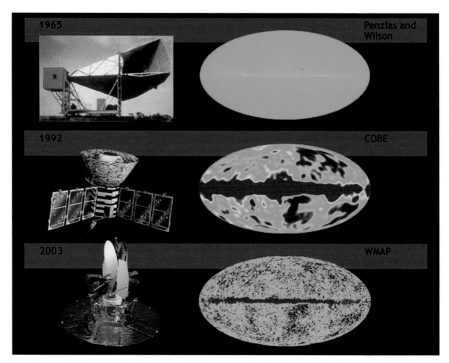

The Cosmic Microwave Background (CMB) was originally discovered by Penzias and Wilson in 1965. It is very difficult to measure from the surface of Earth, so satellites such as COBE, WMAP and most recently PLANCK are used. These measurements unlock all sorts of information about the type of universe we live in. (NASA)

as regular matter does. It also shows something even more mysterious and completely unrelated called dark energy, about which we know next to nothing.

Increasing computer power has given us the ability to make better and better models of how the universe evolved. It is found that models with just regular matter do not work at all well. The universe is very hot at the beginning and so regular matter doesn't settle down into galaxies and stars to begin with as it is continually being disturbed by radiation. Dark matter which is not affected can quickly get on with forming structures that will attract the normal matter once it settles down. So models with dark matter in reproduce the universe we see to a far better resolution, and much quicker than those without.

There are other indications that point the same way, that there is something else out there which is more difficult to explain. Big Bang Nucleosynthesis, the stability of rotating collections of particles, weak lensing and baryon acoustic oscillations are all techniques that indicate extra mass.

What could it be if so?

So, if there is reasonably good evidence that something like dark matter exists, what could it be, and what properties should it have? We can list the properties it must exhibit and so narrow down the field.

- Stability – dark matter must be stable – it can not decay significantly like radioactive elements. We see its effects today, and we need them at the beginning of the universe, so it needs to last at least 13.8 billion years.
- Gravitationally significant – we need it to supply missing gravity so it must interact with gravity. That is after all how we detect it in the first place.
- Transparent – we can not see it, so it must not interact with light. Incidentally dark is a poor choice of name, and transparent might be better, but no one is going to change now!

So – is there anything around that might fit this bill?

MACHOs

There have been suggestions that known objects could provide the solution. MACHOs (Massive Compact Halo Objects), which could be anything from small black holes, neutron stars, brown dwarfs and so on. These are all dark objects – they don't shine like stars so could be overlooked There would need to be a lot of them, and in the right places for this to work, and although a few objects have been detected circling the galaxy, nowhere near enough of them to account for what we see.

WIMPs

At first sight the neutrino seems ideal. It is a very lightweight particle that hardly ever interacts with anything. There are a lot of them about so if we had to dig through our current known particles and come up with a candidate it would be ideal, except for one thing. It travels at the speed of light. This makes it a "hot" particle, and dark matter has shown to be cold both by simulation and observation. A Hot Dark Matter particle would give us a very different sort of universe, with massive structures forming first and then fragmenting, whereas with Cold Dark Matter small structures form first, and then combine. This is what we see happening. So the neutrino is unfortunately not suitable... although maybe it is – see later.

So if it is not anything we have already come across what else could it be? We know our understanding of particle physics is incomplete, and several models have been proposed to help complete it. One of the favourites is call supersymmetry or SUSY for short. This model predicts a complete new set or particles such as squarks, selectrons, photinos and so on. None of these we have any evidence for, but there is a good theoretical framework that means it is attractive. One of the proposed stable SUSY particles is called the neutralino, which would be an ideal candidate for dark matter if it existed.

The Neutralino is one of a class of particles collectively called WIMPs (for Weakly Interacting Massive Particles) and perhaps the leading candidate. Included in the WIMPs are other particles, such as the Sterile Neutrino – a new form of neutrino not so far detected. There are plenty of other theoretical possibilities too – with perhaps less and less confidence. This includes things that come out of string theory, extra dimensional models, variants on the Higgs particle, and so on.

The other perhaps major candidate is the axion (or its SUSY partner the axino). This particle comes out of a different branch of physics and solves some issues which are rather complex to explain within the field of quantum chromodynamics. However besides solving issues in QCD, they would also happen to be a fine dark matter candidate.

After that we descend into all manner of wild possibilities, WIMPZILLA, Q-ball, gravitino, KK graviton, fuzzy CDM, cosmic strings, black hole remnants, have all been postulated, to name but a few.

We have no shortage of volunteers to step forward and assume the mantle of dark matter; although there is one problem – we are unable detect their presence (except by gravity). So how might we go about this?

The Large Hadron Collider (LHC) allows scientists to crash together particles at very high energies. As Einstein's famous equation E=mc² says, matter (m) and energy (E) are related so with enough energy new matter can be created. This can be regular matter or exotic matter, such as dark matter if it exists. The LHC only has a certain energy range, so if dark matter exists beyond the energies available at the LHC it can not be created. (Wikimedia Commons/Maximilien Brice/CERN)

How could we detect it more concretely?

So – of the four ways we know things interact – gravitationally; electromagnetically, and through the weak and strong nuclear forces, we know it must respond to gravity. We also know it doesn't interact with electromagnetism either, else we could see it. More complex reasons also tell us it can't interact with the strong nuclear force either. So that only leaves the weak nuclear force as a possibility. This force is the one that neutrinos were detected through. As its name implies it is not a promising candidate, as its effect is quite limited. However it has been used to detect neutrinos, although even the very biggest detectors struggle to detect more than a handful of neutrinos a day. They are so elusive that even a light year's worth of lead would only stand a 50% chance of stopping an individual neutrino. There are three ways we could detect dark matter directly, popularly known as shake it, make it and break it.

Shake it

Now we know that possible dark matter candidates are most probably going to be even less likely to interact with regular matter than neutrinos, so it is not going to be an easy job to detect them. However many teams have stepped up to the challenge. They build detectors of (in some cases tons) of xenon, or silicon or some other element, hoping to catch the extremely infrequent interaction that might be caused by a dark matter particle. Of course it is not nearly as simple as that. Lots of other things can cause the same sort of trigger in your detector. Cosmic rays, radioactive decay in the surrounding rocks, even background radiation from the materials used to build the detector. So if you are looking for maybe a few events a year, you have to isolate it as much as possible from other sources. Going deep underground helps protect from cosmic rays, protecting the equipment with shields, and making it out of the purest materials you can find are all key parts for hunting these weak signals. As of now – no such signal has been detected – although there have been a couple of false alarms.

Make it

We could also try making it. This is what the Large Hadron Collider (LHC) is good at. If you slam things together hard enough with enough energy, you can make new particles. This is how the Higgs was discovered. Given dark matter's obvious reluctance to interact with normal matter, it would not be detectable in such collisions directly, but it could appear as a missing part in a collision. When all other bits are added up, and if there is something still unaccounted for, this could be dark matter – or something else. We haven't seen such things to date.

Break it

The third main way to try and find it is to use one of the possible deduced properties it might have. It is thought to be its own anti-particle. Therefore if two dark matter particles collide, they will annihilate themselves in a burst of energy. Given how it doesn't like to interact with anything including itself, this would be an extremely rare event. However you can stack the deck somewhat by looking in places where this is more likely. At the centre of the galaxy the dark matter will be much more densely packed, so this is a good place to look for a high energy signal. Unfortunately the centre of the galaxy is packed with

Dwarf galaxies, such as the Fornax dwarf galaxy pictured here, are popular places to look for signatures of dark matter. They are big enough that the centre would concentrate dark matter into a hopefully detectable signal, but not complicated enough that it would be difficult to confuse the signal with other events. This does all depend on the supposition that dark matter will interact with itself in sufficient quantities to make such a signal. (ESO/Digitized Sky Survey 2)

lots of other exotic things, including a super-massive black hole. So, almost any energy signal detectable from there can have an equally plausible explanation. Thus it is more popular to look in dwarf galaxies which are a little less messy to untangle. Despite a few false alarms, still nothing tangible at present.

What are the alternatives?

So – going back to the start of the article, and given we haven't managed to detect it yet, what if we are wrong? A dark matter particle seems a good answer, being able to fit nearly all the data. Should we consider other options? Well, being good scientists, of course we should.

One of the first candidates is something called Modified Newtonian Dynamics (MOND). This involves tinkering with some of the basic equations from Newton, in particular maybe the second most famous equation in physics, $F = ma$. The suggestion is that this is changed to be something more complex, $F = m\mu\left(\frac{a}{a_0}\right)a$ which becomes $F = ma$, except that when a is very, very small then good things happen. This model fits the rotation of galaxies extremely well, provided the right constants are chosen. On its own it struggles to explain some other phenomena, and requires changes to incorporate into other theories – it violates the conservation of momentum law for instance. Enhanced models such as TeVeS and BIMOND have been put forward to correct for this, but have issues of their own.

Conclusion

So what do we make of all this? It is still anyone's game, as there have been no solid detections of any of the theories. As I said in the beginning, most astronomers tend to side with dark matter, but dark matter detectors are still coming up empty. This may just be because it hides itself extremely well, so well in fact that we will only ever be able to detect it by its gravitational signature. On the other hand, it may be something completely outside our imagination at this point. Keep watching this space!

Eclipsing Binaries

Tracie Heywood

The story of eclipsing binary stars begins in the year 1782. John Goodricke, a young man who lived in York, had been investigating the brightness changes of the star Algol (Beta Persei). It had long been known that Algol would occasionally dip in brightness by more than a magnitude. The Italian astronomer Geminiano Montanari (Figure 1), in 1667, was the first European observer to report its changes, but it is highly probable that the brightness changes had been noticed by Chinese and Arabic astronomers many centuries earlier.

Goodricke's careful monitoring of the brightness of Algol indicated that the dips in brightness occurred at regular intervals. He proposed that the dips in brightness were due to a "dark object" that was in orbit around and periodically passing in front of Algol, as seen from our line of sight. The capture of spectra of Algol in 1888 and 1889 by the German astrophysicist Hermann Carl Vogel revealed that the spectral lines of the star did show Doppler shifts. The pattern of changes revealed that when approaching eclipse, the star was receding from us and when emerging from eclipse it was approaching us, thus confirming the orbital motion. Algol was revealed to be a blue-white main sequence star, but the exact nature of the secondary object remained unclear because its spectral lines were not prominent enough to be recorded. The detection in 1910 of a (shallow) eclipse of the secondary object showed that this object was not fully dark. It wasn't until 1954, however, that the spectrum of the secondary object

Figure 1. Portrait of Geminiano Montanari (1633–1687), the first European to report changes in brightness of the star Algol. (Geminiano Montanari/L'Astrologia Convinta di Falso/Wikimedia Commons)

was captured, revealing it to be an orange subgiant star. Thus, the Algol system includes two stars. The deeper (primary) eclipses occur when the brighter blue-white star is hidden by the fainter orange star; the shallow (secondary) eclipses are due to the fainter orange star being hidden by the blue-white star.

Beta (β) Lyrae

The brightness variations of Beta Lyrae were discovered by Goodricke in 1784. It was observed to have eclipses approximately every 6.5 days, but closer inspection showed that deep and shallow eclipses alternated, the deeper eclipse being due to the eclipse of the brighter star in the system by the fainter star. The shallower eclipse is a secondary eclipse caused by the brighter star passing in front of the fainter star. Both eclipses are readily observable because the two stars are more closely matched in brightness than those in the Algol system. Consequently, the period of Beta Lyrae is listed as approximately 12.9 days. It was also observed that the brightness of Beta Lyrae is not constant between eclipses, but continues to change, albeit much more slowly. This is a consequence of the two stars having gravitationally distorted each other, becoming more egg-shaped than spherical. As a result, the overall brightness of the Beta Lyrae system is greatest midway between eclipses when we see these "eggs" side on.

Types of Eclipsing Binary

Algol and Beta Lyrae define two of the three main types of eclipsing binary. As we have already seen, in Algol type systems, the brightness is constant between eclipses, whereas in Beta Lyrae type systems, the brightness continues to change between eclipses. The third type of eclipsing binary is defined by the star W Ursae Majoris. In these systems, the two stars are so close together that they are effectively in contact and, as a result, there is no gap between eclipses.

Eclipses can be partial or total, depending on the relative sizes of the two stars and how closely aligned their orbital plane is with our line of sight. The two types are quite easy to distinguish. Light curves for total eclipses, such as that of U Cephei (Figure 2), have flat bottoms. These feature a spell of constant brightness near mid-eclipse while the brighter star is completely hidden. Light curves for partial eclipses, such as that for RZ Cassiopeiae (Figure 3), are more "V-shaped". In these systems, the direction of brightness change reverses as soon as minimum brightness has been reached.

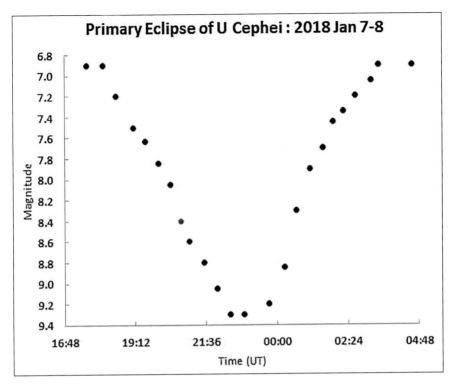

Figure 2. Light curve for the primary eclipse of U Cephei during the night of 7/8 January 2018, recorded by Tracie Heywood using 11x80 binoculars. This eclipse is flat-bottomed, indicating that it is a total eclipse. (Tracie Heywood)

Discrepancies

Having determined both the orbital period of the system and the mid-time of one eclipse, it ought to be straightforward to predict the times of future eclipses. Over the decades, however, it has become clear that the observed times of eclipses gradually diverge from such predictions. Why should this be? Surely star systems remain unchanging for centuries or even millennia?

Could it be a problem with the accuracy of the later observations?

Although some eclipse observations are made by professional astronomers, the longer-term monitoring of eclipses has been very dependent on visual observations made by amateur astronomers.

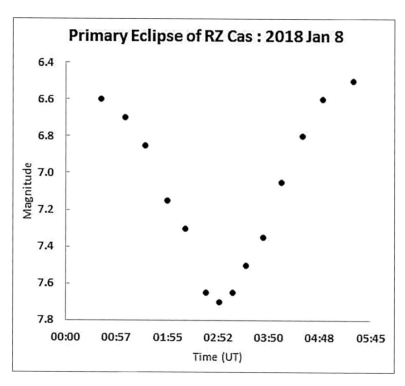

Figure 3. Light curve for the primary eclipse of RZ Cassiopeiae during the morning of 2018 January 8, recorded by Tracie Heywood using 11x80 binoculars. This eclipse is "V-shaped", indicating that it is a partial eclipse. (Tracie Heywood)

Visual observing is the most convenient method of observing and until recent decades was the only option available. However, a number of factors can affect its accuracy, including the presence of cloud, haze or moonlight and the glare from street lighting. These factors will often affect some parts of the sky more than others, a fact that can be significant because during an eclipse lasting, say, 9 hours, the Earth's rotation will move the star across the sky.

The length of eclipses introduces additional factors. The whole of the eclipse may not fit into the period of darkness. Some eclipses will start or end in daylight or twilight. Adding in the possibility that parts of an eclipse may also be missed due to cloud means that eclipse light curves can often be incomplete. There are various methods to circumvent this, but all have their limitations. Gaps can, for example, be filled by using observations made on other nights or by other observers, but these may have been made under different sky conditions.

The effect of such issues can be seen in the light curve for Beta Lyrae (Figure 4), which is based on observations made during 2016 by members of the Variable Star Section of the Society for Popular Astronomy. With eclipses of Beta Lyrae being too long to fit into a single night, it becomes a case of making a brightness estimate on each clear night during the year and then merging these to produce a single light curve showing one cycle of variation. The deeper primary eclipse and shallower secondary eclipse can clearly be seen in the light curve, but the variation of sky conditions from night to night and location to location has led to an increase in the scatter in the observations.

When other factors are added in, including the colour sensitivity of each observer's eyes and the risk of observers being unduly influenced by the trend in their earlier observations, it might seem feasible that the discrepancies between predictions and observations could simply be due to the limitations of visual observing. However, the discrepancies observed, being in many cases measured in hours and, in some cases, days, are far too large to allow such an explanation.

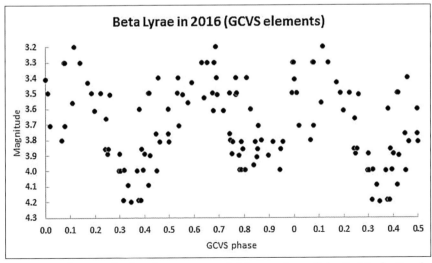

Figure 4. Light curve for Beta Lyrae, as recorded by members of the Society for Popular Astronomy during 2016. (Society for Popular Astronomy)

Could the problem be with the accuracy of the original observations?

It would only require there to have been a small error in the original listed orbital period to produce measurable discrepancies at a later date. With the periods of eclipsing variables typically being no more than a couple of weeks, many tens of orbits are completed in each year and many hundreds are completed per decade. Small errors, of the order of seconds, could accumulate over hundreds or thousands of orbits to produce discrepancies in eclipse times that are readily apparent to later visual observers.

This cannot be the explanation, however, because more recently determined measurements of the period when extrapolated back in time are well out of step with the originally observed eclipses. The discrepancies are clearly real and it must be the orbital period itself that is changing over time. Something is physically changing out there.

What might cause the period of an eclipsing binary to change?

There are two main possibilities:

Mass loss/transfer:

Most eclipsing binaries have short orbital periods, measured in hours or days. Periods longer than a fortnight are quite rare. The longer the orbital period, the further apart the stars will be in space and the lower the likelihood that they will align to produce eclipses as seen from our line of sight.

When you consider that in our solar system, Mercury takes 88 days to orbit the Sun, you get an idea as to how close the stars must be in eclipsing binary systems.

Stellar evolution theory tells us that the more massive star in a binary system will evolve fastest. In most cases it will evolve into a red giant and in doing so, its outer layers expand and become less securely attached gravitationally. The less massive star may then capture some of this escaping material. The changes in the relative masses of the two stars causes the orbital period to change. Consequently, the eclipses slip out of line with predictions based on the earlier orbital period. Even a small change of a few seconds in the period will accumulate over the many orbits each year to become a readily detectable discrepancy in the eclipse timings.

Gravitational effect of third star in the system:
Although we are only directly aware of two stars in the system (due to the eclipses), this doesn't rule out the possibility of a third star also being present. Although not involved in the eclipses, this third star can gravitationally distort the orbit of the eclipsing pair. When this third star is on one side of its orbit it will cause the eclipses to occur early; when it is on the other side it will cause eclipses to occur late.

O-C Diagrams
Astronomers monitor the long-term period changes by plotting O-C diagrams, showing the discrepancies between the Observed and Calculated times of eclipses. In such diagrams, the presence of a third star in a circular orbit would produce a fairly sinusoidal pattern, with the eclipses occurring early for a number of years and then late for a similar number of years. The time taken to complete the cycle would match the orbital period of the third star. In contrast, mass transfer produces more irregular changes, dependent on the rate of transfer, but over time the discrepancies increasingly move in a particular direction.

In some eclipsing binaries, a combination of the two factors will be present. This will result in the pattern of discrepancies being somewhat more complicated.

Improving the Accuracy of Observations
Early eclipse observations were made visually. This led to a fair amount of scatter in O-C diagrams and consequently it was not always straightforward to recognize trends.

Nowadays, professional astronomers rarely accept eclipse timings made visually and instead require the more precise photometric measures that can be obtained using CCD cameras or DSLR cameras. Making brightness measurements in this way is somewhat more challenging but, provided that sufficient care is taken, the accuracy of the brightness measurements can exceed that of visual observations by a factor of five or more. An added benefit is that there is no risk of the "observer" being unduly influenced by previous measurements. Extra work is involved because, whereas our eyes can readily see the brightness of a star and ignore any light from surrounding stars and the sky background, the detectors in these cameras capture everything, including any background "noise" from the camera itself. Thus, additional measures need to be taken with each observation of the star in order to subtract out the light of the sky background and the "noise" of the detection system.

Figure 5 shows the light curve for an eclipse of the star AD Andromedae recorded by David Smith during the night of 29/30 January 2018. David recorded the brightness changes using a CCD camera, making a measurement every 20 seconds.

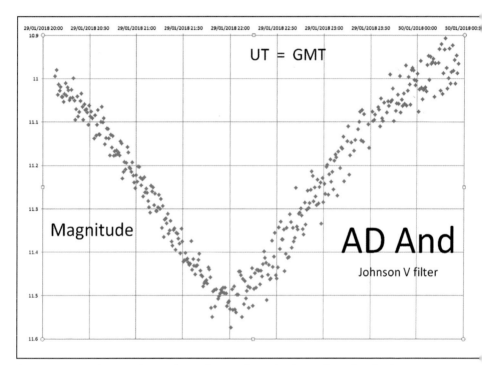

Figure 5. Light curve for an eclipse of AD Andromedae during the night of 29/30 January 2018, recorded by David Smith using a 90mm refractor (William Optics Megrez 90) with a 0.8x focal reducer and an Atik 460EX mono CCD camera, all mounted on an NEQ6 mount. Note that the scatter increases slightly during the course of the eclipse due to the star becoming lower in the sky. (David Smith)

Additional Features of Eclipses

When a number of eclipses of a star have been observed, additional features may be noticed.

Asymmetrical Eclipses

Some eclipses are revealed to have asymmetrical light curves. In visual observing there can sometimes be a psychological explanation for this – the observer unconsciously rushing the completion of the eclipse so that they can go to bed. However, DSLR and CCD based observations show that for many stars the effect is real.

This asymmetry would suggest that there is something unusual about one or both of the stars.

There are two main explanations, one being that the surface brightness of the star being eclipsed is non-uniform. This can affect the rate at which the brightness drops on entering eclipse and rises on leaving eclipse. The most common cause is the presence of star-spots, much larger than those seen on our Sun.

The second possibility is that one or both of the stars is itself intrinsically variable in brightness over short timescales, these being comparable with the duration of the eclipse.

Variations in the depth of eclipses

One or more of the stars may be variable in brightness over longer timescales. If this is the case, then the overall brightness of the system outside of eclipse may change from eclipse to eclipse and more noticeably, some eclipses may be deeper than others.

DSLR/CCD brightness measurements can also bring other "features" into view, including limb darkening and reflection effects. Some of these may show up in visual observing but, in most cases, they are too small to clearly stand out visually.

Making Your Own Observations

Although professional astronomers rarely accept visual timings of eclipses, it can still be a rewarding experience to observe an eclipse visually, seeing the star fade in brightness over a number of hours and then brighten again.

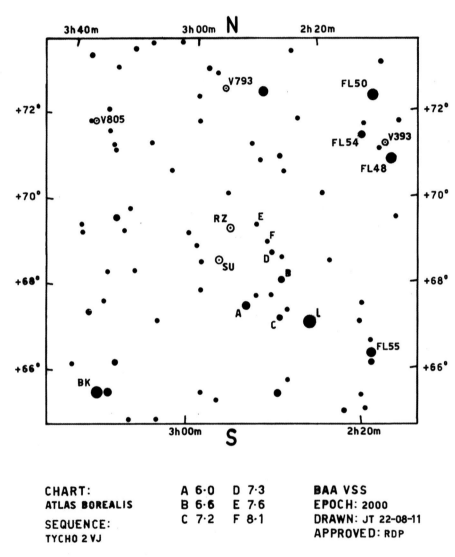

Figure 6. Finder chart for RZ Cassiopeiae, provided by the Variable Star Section of the British Astronomical Association. (BAA)

If you would like to make your own observations of an eclipsing binary, a good candidate is RZ Cassiopeiae, which can be monitored using 50mm binoculars or a small telescope. RZ Cas has eclipses that are not only frequent (approximately every 29 hours), but are also deep (fading by over a magnitude) and are reasonably short (lasting for 4.8 hours, but with most of the "action" during the middle 3 hours). In addition, Cassiopeia is circumpolar from the UK and so eclipses can be observed throughout the year.

The accompanying chart (Figure 6) shows the location of RZ Cas. It also labels some nearby stars (A to F) that are unchanging in brightness. The brightness changes in RZ Cas can be followed by comparing it with these stars.

Having looked up the times of upcoming eclipses and located RZ Cas, make an estimate of its brightness every 20–30 minutes by comparing it with the lettered comparison stars. At the end plot a light curve showing your results. Don't worry if the mid-time of your eclipse differs from the predicted value. Predictions are often calculated from orbital periods derived years ago. When the orbital period is constantly changing, it is impossible to predict eclipse times in advance with 100% accuracy.

Predictions

In order to predict the times of upcoming eclipses, all that you need to know is the orbital period of the system and the mid-time of a previous eclipse. Actually, it is a bit more complicated than that because you need to get involved with Julian Dates. Fortunately, there are websites around which have already done the hard work for you.

If you want to find out which stars are due to have eclipses tonight, the best source is the website of the British Astronomical Association (BAA). This gives predicted start, mid-eclipse and end times to the nearest hour. The site is at **http://www.britastro.org/vss/dpredict.html**

If you want to identify upcoming eclipses of a particular star, the best source is the predictions provided by the Mt. Suhora Astronomical Observatory. For RZ Cassiopeiae, select "Cas" from the constellation list and then select "RZ". Then click on "Current Minima and Phase" and page down to see the predicted mid-times for the next six eclipses. The site is at **http://www.as.up.krakow. pl/o-c/index.php3**

A Perspective on the Aboriginal View of the World

Greg Quicke

'Isn't it amazing how the Aboriginal people knew about the stars and used them as navigational tools, seasonal indicators and incorporated them all throughout their lore and culture?' I am often asked. Well, 'yes' and 'no' is my answer. Yes, because it is always amazing when you can tune into everything that is around you and see the connections between seemingly unrelated things. And 'no', because these people have lived under the stars for a good many thousand generations and have watched these cycles until they have become a part of who and what they are.

I know this very well because of my own learning process. People often ask me how I learned and how I know so many of the stars and their names. My answer is that I learned them the same way everyone learns anything. I learned them one at a time. I will often ask the kids, 'On your first day at school, did you know the names of all of the kids in your class?' Usually, the answer is 'no'. I bet that by the end of the year, you know all of their names and all of their characters and a hundred other things about them. And you have learned these things and these people one at a time. This is the same way you learn the stars.

Now, imagine being born into a culture that lived under the stars for generations, where you grew up listening to the stories set to the very stars you lived under every night. These stars would become like old friends, and of course you would know them as well as you know your own family. So, have I made an academic or formal study of the Aboriginal way of looking at the stars and the sky? No, I haven't, because that is not the way that I learn. Instead, I have watched and listened, hearing some of the stories of the elders and the story keepers enough times that small parts of them have become a part of my psyche. To be clear, I write about the Aboriginal perspective of the night sky through no authority other than my own experience of half a lifetime living in a place and around the original people of the land where the natural ways of life are still quite normal.

The wild Kimberley region of Western Australia is home to more than 30 different Aboriginal tribes, all holding a particular aspect of the law and culture, each preserving the integrity of places that are treated as live entities connected with the land, the sea and the sky. (Greg Quicke)

Aboriginal people have watched this sunset from the cliffs of Gantheaume Point for tens of thousands of years. Here, Greg Quicke does the same, waiting for the first glimpse of Venus in a quickly darkening tropical sky. (Leon Mead)

In making a film about ancient songlines for the BBC's Stargazing Live with Dr Noel Nannup of the Noongar people of southern Western Australia, I learned from Noel that the oral tradition, the spoken word way of passing down stories and knowledge, is a very precise one. Every pronunciation, every inflection, every tone of voice must be perfect before the highly honoured mantle of story keeper is passed on. It is the same with the equally revered keepers of the songs and the keepers of the law. To my mind, this makes this knowledge incredibly reliable over vast periods of time.

Kimberley Tides

Where I live in the wild Kimberley region in the far north of Western Australia, adjusting your life around the natural cycles is one that can make the difference between life and death. My good friend Narnie Howard, with his generations of aboriginal knowledge, deeply etched into his psyche, had to convince a group of visitors that it was time to stop fishing as we had to cross back over a

This powerful and sustained positively charged lightning bolt lit up a 6 hour storm high up on the wild Kimberley coast. Making connections from planet to planet, from sea to sky, from culture to culture. (Greg Quicke)

tidal creek before the tide turned. 'Just one more cast' on the seemingly calm creek bank turned into a shouting match of 'We have to go now!' Sure enough, we found ourselves chest deep in a powerful current threatening to drag us miles up a crocodile-infested Kimberley creek. We eventually dragged dogs and children and piles of fishing gear to the other side where everyone collapsed in a wet and muddy heap. Looking at each other with exhausted gratitude we had gained a fresh respect for the country and were all eyes wide open with respect for Narnie. Experience is a great teacher.

I learned first hand about Broome's massive 10 metre tides working as a pearl shell diver. Instead of working the weekdays and having the weekends off, we worked 10 days over the neap tides and had four days off over the springs. The tides are governed by the Moon. Spring tides are big. They happen when there is an alignment between the Earth, the Sun and the Moon. This happens at New Moon when the Moon comes between us and the Sun, and it happens again at Full Moon when the Moon is opposite the Sun from our viewpoint here on Earth.

I learned to watch this cycle of the Moon going around the Earth and to gradually put the pieces of the jigsaw puzzle together with the tides. I watched the Moon being high in the sky and the tide being full soon after. I watched the Moon on either the western or the eastern horizon and noticed the tide being out. Although this was confusing for me in the beginning, I did what the Aboriginal people had been doing with the tides for many thousands of years. I watched them, and then I watched them again. I started to see a pattern emerging. I thought I had it and then the tide messed me up by happening four times in a day instead of the two tides that I was expecting. I watched some more. If the Moon pulls the tide up underneath itself when it is high in the sky, why is there a high tide when the Moon is below the horizon? That one threw me for a long time.

I thought about the centre of gravity between the Earth and the Moon not being in the centre of the Earth and started to appreciate that the Earth and the Moon wobble around a common centre of gravity. In that moment of realisation, I began to sense the movement, feeling the dance of the Earth and the Moon with each other. I realised that this dance resulted in the tide 'throwing out' on the opposite side to the Moon and that the Earth turned around once a day within these two bulges of water, one drawn up under the Moon's position and the other thrown out on the other side of the Earth. The resulting two high tides and two low tides in a day then made themselves known to me. I started to look at the Moon's phase and its position in relation to the Sun to figure out where the tide was going to be that day, how high it would be and at what time. I started to get really good at it, simply because I was using that information every day. I made plenty of mistakes and each mistake fine-tuned my knowing until, after a while, I did not need a tide chart anymore.

The Aboriginal people, who I know and call my friends and family, put this into practice every day in the most natural way. The wind will change and they look up, knowing that the tide has turned. The little bird that runs along the beach cries out and tells them the same thing. How do they know this? They know it because it has happened again and again and through thousands of years and countless generations.

Watching the night by night changes of planetary conjunctions, like this one with Mercury and Saturn in November 2017, is a natural part of life for the Kimberley Aborigines. (Greg Quicke)

Desert People

On the edge of the Great Sandy Desert with the Walmajarri people, I found a clear area to do some stargazing. No sooner had I started to unload and set up a serious fleet of 8-, 10- and 12-inch telescopes when the old ladies started turning up to sit in the dirt and watch. 'You picked a good place,' they told me. 'This is women's ground and everyone is welcome here. If you picked a different place we would not have been allowed to come.' This felt immediately good. Soon we were surrounded by a couple of hundred children; the men came too and at least a hundred dogs. Some young men pointed at a particular part of the Milky Way and told me that was where they had come from. They pointed at another part of the sky and told me that was where they were going next. There was precision in their pointing and in their knowing. Then a poke in the ribs let the speaker know he had said enough. Some of the sky knowledge is public and can be talked about and some of it is only for the initiated few.

This 10 inch f/5.6 Dobsonian telescope has done excellent service throughout the Western Australian outback for over 30 years. Thousands of people including many aboriginal communities have had their first views of the Moon, Saturn and many other wonders of the heavens through this worthy workhorse of the skies. (Leon Mead)

The old ladies asked me to come and sit. 'We've got stories about the stars too you know.' I sat waiting with great anticipation. 'We can't tell you though. They are secret women's business.' I sank a little although with the greatest respect for their integrity. 'We can sing you the songs though!' they beamed. And I beamed back. The songs and their voices seemed to come from deep underground, connecting us with the Earth, vibrating all through my body then out the top of my head to connect me with the sky. I have goose bumps writing about it. This was one of the most amazing nights of my life, sitting in the dark with only the magnificent stars of the Milky Way to light up the faces and the smiles and the bright white teeth.

These desert people were just as interested in the perspectives I offered, laughing and giggling and sharing with each other what they were seeing in the telescopes with close-ups of the Moon, the rings of Saturn and the star clusters that they so beautifully depicted in their art. The Aboriginal people do see things in a different way. Artworks of the land, as accurate as any aerial photograph, reveal an ability, and a knowing of their country, that is beyond western understanding. Noticing a hundred things about a landscape that a white fellah sees as a barren wasteland is what has allowed them to thrive in the Australian outback for many thousands of years.

Saltwater People

In walking the ancient songlines in the country north of Broome with the Goolarabooloo, the Nyul Nyul, the Bardi, the Djabbir Djabbir, the Nimanburu and the Yawuru peoples, I listened and learned, more through osmosis rather than direct teaching. My old friend Lulu, who has since passed into the Bugarrigarra, or dream time, taught me to notice my environment by asking questions about the different places we walked through. 'How did it make you feel when we camped at this place?' And 'How did it make you feel when we camped at that place?' And 'When we stopped at that other place for lunch and made clap sticks, how did you feel?' As white fellahs, we think of real estate when aboriginal people talk about country and land. I have learned through Lulu and his family on the Lurujarri Heritage Trail that each place in the land has character, it has a unique and individual life that can be tuned into if we will only take the time to notice and to question. Lulu taught me that as you treat the land, so the land will treat you back. 'You have to help us keep this land alive,' he told me. 'It doesn't matter if you are black, white or brindle. If you walk this land with the right things in your heart, this country will smell you, it will come to know you and it will give you what you need.' 'If you come onto country with the wrong things in your heart, this country will smack you, trying to teach you how to live and how to think and how to do the right thing by everyone and everything around you.'

My Goolarabooloo family taught me and showed me and have sung me the story of the Emu Man in the sky. According to aboriginal lore, Marala is the creator god who made all of the country. He made the trees and the ground and the kangaroos. He made the reefs and the fish that swim around on the reefs to feed the people. While he was doing this making, as creator gods do, he taught the people right from wrong. He walked all over the Broome coastline, leaving his footprints behind. Today we call the footprints of Marala dinosaur tracks. The Broome coast sandstone is littered with thousands of footprints from 21 different species of dinosaurs, from the elephant-like tracks of the giant Sauropods to the 3-toed 'T Rex type Theropods. The 3-toed 'Emu Man' left his last set of tracks at Gantheaume Point, Minyirr, before going up into the sky. We can see him today in the dark dust lanes of the Milky Way. His head is the Coal Sack Nebula, nestled into the Southern Cross. His neck runs past the Pointers in Centaurus, while his body and the frills of his skirt swell out around the Scorpius/Sagittarius region. His legs spilt as they run out towards the constellation of Aquila, the Eagle. Once you see Marala, the Emu Man, stretching most of the way across a dark moonless Kimberley night sky, you will wonder how you failed to see him before. While you are watching him, remember that he is also watching you, to make sure you know right from wrong.

Different stories are told around the campfire for different locations as the people move through the land. The stories and the songs and the laws pay homage to the features and the gifts that each living place has to offer with a recognising of the links between the land, the sea and the sky. A particular set of rocks may correspond with a set of stars in the sky. Certain trees are living entities connected with their contemporaries in the sky. Waiting and watching for particular stars appearing at particular times of the year, corresponding with different foods being ripe, different fish being fat, and indicating the time to hunt goanna. Broome's own band, the Pigram Brothers, sing their 'May Song' that

The 3-toed dinosaur tracks of Marala, the Emu Man, preserved in the Broome sandstone. He left his last set of tracks at Gantheaume Point, Minyirr, before going up into the sky where he still watches all of us to make sure we know right from wrong. (Greg Quicke)

tells us that 'the scorpion is rising in the east. Orion, well he's just gone to sleep (in the west). Corvus crow, watch him diving down (from up on high), Simpson beach seems so far away from town.' It's a song about salmon fishing in the month of May when Scorpius is rising, Orion is setting and Corvus is overhead. The swallow-tailed kites with their tail feathers shaped like the salmon turn up at this time too, as another indicator of the intimate connections between land, sea and sky.

Perspectives

Western science brings us amazing perspectives on our physical world as we learn more and more how to measure and interpret the things we see around us. The Aboriginal people have taught me to look for the things that are not

The dark shape of the Coal Sack Nebula, also being the head of Marala, the Emu Man, can just be glimpsed low over the south eastern horizon, pushing the Southern Cross up into the sky. The mistiness of the incredibly rich Milky Way forms a beautiful backdrop to this picture, with the Eta Carinae star-birthing region sitting just above centre of image. The Large and Small Magellanic Clouds seen to the right combine with two meteors streaking across the sky to add intrigue to this wonderful view. Living under a sky like this there is no surprise that the first Australians are well and truly tuned in. (Leon Mead)

so obvious, for the things that require a trained eye to see and a trusting of gut feeling to guide and a questioning mind to put together. This is also the scientific method that has been honed by the passage of time, by the watching of cycles, by the making of mistakes and by the beautiful art of observation.

Each time I consciously witness the Earth turning in space and travelling through space on its very real journey around the Sun, there is something that deepens in my understanding of this journey. There are cycles within cycles within cycles and living by these cycles is something that we all do whether we have an awareness of them or not. The Earth turning and going around the Sun, the Moon going around the Earth, each of the planets interacting with the Earth in particular ways to bring them into and out of our night time sky, are all things that we can gain a spatial awareness of. After all, this is our local environment of space. The aboriginal people have taught me and continue to teach me to observe my earthly and celestial environment, and for this I am eternally grateful.

The First Known Black Hole

David M. Harland

Nowadays the term black hole is in common usage, such as to highlight an unfortunate shortfall in a budget. Many people are also aware of the origin of the term in astrophysics. But there was a time when the notion of a black hole in space was new and distinctly heretical. Even specialists never expected that astronomers would actually find such things. Indeed, those most familiar with the concept were ardent that black holes simply couldn't exist! Albert Einstein insisted that despite the mathematical logic, Mother Nature would never allow such an object. As the leading theorist Kip Thorne of the California Institute of Technology later reflected, they were "more at home in the realms of science fiction and ancient myth than in the real universe."

This article reviews the theoretical prediction of this amazing phenomenon and the momentous discovery that such things really do exist.

When Isaac Newton published his theory of Universal Gravitation in 1687 what we call physics was known as natural philosophy, so he entitled his tome *Philosophiae Naturalis Principia Mathematica* (Mathematical Principles of Natural Philosophy).

Edmund Halley, an astronomer and friend of Newton, then used the theory of gravity to predict that a comet which had appeared in 1531, 1607 and 1682 would return in late-1758. When this occurred, spectacularly demonstrating the power of mathematics, Halley was dead but the object of his prediction became known as Halley's Comet.

In 1767 John Michell suggested on a statistical basis that most of the stars that are close together in the sky are gravitationally bound together, as 'double stars'. This proposition was received with considerable scepticism. But at that time astronomers had no interest in the stars beyond their role as a background to measure the motions of solar system objects.

While 'sweeping' the sky in 1783 for interesting objects, William Herschel, who knew of Michell's idea and routinely took note of double stars, identified

40 Eridani as a *triple* star system. It is a binary in which one of the two is itself a binary. There was no way for Herschel to know that the faint white star in the 'inner' system was of a type that would become known as a 'white dwarf', but he was the first to see one.

After a decade of astrometric measurements of the positions of Sirius and Procyon in the sky, Friedrich Wilhelm Bessel realised in 1844 that both were 'wobbling' as they pursued their 'proper motion' tracks. From this, he inferred that they possessed unseen companions. From the 50 year periodic motion of Sirius it was evident the companion had roughly the same mass as the Sun, so that star was not only very small but also very dense. What kind of object was able to perturb a star and yet remain invisible?

In 1862, Alvan Graham Clark inspected Sirius to perform a test of the 18-inch refractor (the largest lens in the world at that time) which was destined for the Dearborn Observatory of the University of Chicago, and was astonished to *see* the companion as an 11th-magnitude star almost overwhelmed by the glare of its primary. John Martin Schaeberle saw Procyon's companion in 1896 using the 36-inch refractor of the Lick Observatory.

These intrinsically faint companions were presumed to be slowly cooling and very old stars, because at that time stars were believed to start out as large luminous bodies and then contract slowly and cool over aeons. Walter Sydney Adams at the Mount Wilson Observatory isolated Sirius B's spectrum in 1914 and eventually identified highly ionised lines that implied a temperature of 10,000 K. To be both hot *and* faint, the star had to be physically small; no more than 20,000 km in diameter. By some process, the mass of the star had been compressed into a volume no larger than that of Earth.

In 1926 Ralph H. Fowler in Cambridge investigated this extremely dense state. In a star, the relentless compression of gravity is balanced by the pressure in the core. He calculated that when the core was overwhelmed, gravity would squeeze it to a density of 10^9 g/cm^3, collapsing the atoms sufficiently to leave isolated nuclei in what was essentially a gas of electrons. Further collapse was inhibited by the Exclusion Principle of electrons discovered by Wolfgang Pauli in 1925, which manifests itself as the pressure of this 'degenerate' gas. As the core of a red giant star becomes a white dwarf star, the release of gravitational energy ejects the envelope to create a planetary nebulae (which solved another mystery).

While sailing to England in 1930 to work for his doctorate under Fowler's supervision, Subrahmanyan Chandrasekhar, a physics graduate from Madras University in India, calculated there was an upper limit of 1.44 solar masses for a white dwarf. It followed that as a massive star was nearing the end of its life, if its core exceeded this mass when it collapsed, the pressure of the degenerate electrons would not be able to resist further collapse. Of course, the star may have started its life much more massive than this, because it would inevitably shed material as it evolved.

At a Royal Astronomical Society meeting in 1935 the leading theorist on stellar physics, Arthur Stanley Eddington rejected 'Chandrasekhar's limit' for the reason that it implied a star could cut itself off from surrounding space and essentially cease to exist, which he considered to be a ludicrous proposition.

Armed with his doctorate, Chandrasekhar accepted a post at the University of Chicago in 1936, where he established himself as the new leading theorist on stellar structure.

Next, in 1934 Walter Baade and Fritz Zwicky at the California Institute of Technology, seizing on the discovery by James Chadwick the previous year of the neutron as an electrically neutral counterpart to the proton, speculated that if a star that was too massive to become a white dwarf collapsed, the electrons would be obliged to combine with the protons to form a neutron gas. This was significant because neutrons, like electrons, obey the Exclusion Principle, and tend to resist compression.

In 1939 J. Robert Oppenheimer and George Michael Volkoff, a graduate student, calculated how the 'degenerate neutron pressure' would resist further gravitational collapse and result in the production of a 'neutron star' which was only a few kilometres in diameter. The tremendous density and gravity of such an object required them to incorporate General Relativity into their description. Astronomers were not very interested. They reckoned that massive stars would shed sufficient material as they evolved to be able to retire as white dwarfs, so they did not really expect neutron stars to exist. Hartland S. Snyder, another of Oppenheimer's students, realised that if the collapsing core of a star exceeded about 3 solar masses, then the neutron pressure would not be able to prevent a runaway collapse in which the star vanished from space.

In 1916, a year after Albert Einstein announced General Relativity, Karl Schwarzschild in Germany developed the first full solution to the equations of the theory. In thinking about the prediction that light would be deflected when passing close to a gravitating mass, he realised that if a star had a critical ratio of mass to radius then it would *warp* space so severely that not even light could escape. This became known as the 'Schwarzschild radius'. The term 'dark star' was coined, but because the masses of known stars were not confined within such small volumes the idea was dismissed as having no physical significance.

In fact, this was not an entirely new concept.

In 1783 John Michell, the man who had suggested the existence of binary stars, argued that if light was a stream of 'corpuscles' as Newton believed, then by the law of gravitation there must be a certain mass at which a star would be able to prevent light from escaping, rendering it invisible. Several years later, French polymath Pierre-Simon Laplace similarly reasoned that if a collapsing cloud of interstellar gas were sufficiently massive it would produce an invisible *corps obscur* ('dark body') by inhibiting light from escaping. Although strictly speaking this was not the same as the Einsteinian concept, it demonstrated that even with 'classical' physics it was theoretically possible for such an object to exist.

The properties of such compact objects were later studied in detail, but it was reckoned that, in practice, the process of collapse would be disrupted by turbulence that would prevent the *singularity* from forming; it would result in an irregular explosion. Conversely, in 1965 Roger Penrose at the University of London proved that a runaway collapse *must* end in a singularity.

Attending a conference in New York in 1967, John Archibald Wheeler of Princeton coined the term 'black hole'. He selected this name because he felt the singularity was essentially a 'hole' in the fabric of space. The critical radius was labelled the 'event horizon' because no events beyond it were visible from outside. In contrast to the collapsed star that Laplace had envisaged, the event horizon of a black hole is not a physical surface, it is simply a surface in space. The radius of the event horizon is determined by the mass of the singularity at its centre, whose density is infinite.

Roger Penrose of Oxford in England said that Mother Nature imposed the event horizon as an act of 'cosmic censorship' to hide the *flaw* from the rest of the universe.

Just as the existence of white dwarfs had been established by finding them in the binary systems of Sirius and Procyon, in the mid-1960s people started to wonder whether it would be possible to infer the presence of a black hole in a binary system by the manner in which it perturbed its companion star.

At the Physical Technical Institute in Leningrad, Russia, Yakov Borisovich Zel'dovich gave astronomy graduate Oktay H. Guseynov the task of searching through star catalogues to identify stars that might be 'wiggling' as they traced their proper motions across the sky, in the hope of finding an example where the only explanation would be an unseen companion with a mass requiring it to be a black hole. This strategy had been suggested by John Michell in the 18th century, when he envisaged there might be stars so massive that they prevented light from escaping. The necessary observations had been impracticable at that time. By 1966 Guseynov had identified five potential candidate stars. Applying the same rationale, Kip Thorne and Virginia Trimble, an astronomy postgrad at the California Institute of Technology, added another eight candidates. But this research led nowhere.

Another prospect was to examine binary systems for which the plane of the orbit produced eclipses. If a system held a black hole, the radial velocity cycle of the visible star would show a significant motion. By calculating the masses, it would be possible to prove the presence of a black hole. So the strategy was to seek evidence of an invisible massive secondary.

Epsilon Aurigae attracted a lot of interest in this regard. Observations by a number of observers over the years, particularly Hans Ludendorff early in the 20th century, showed it to be an eclipsing binary with a period of 27 years and an unusual flat-bottomed minimum in the light curve. Over the years, a variety of explanations were offered, usually involving a disk of material that masked or reflected light as necessary. As the primary seems to be a supergiant and the secondary is also massive, when it started to look as if black holes might really exist this system was considered. Although the case for a black hole was ruled out, the system remains somewhat mysterious to this day.

What really ignited the hunt for black holes was an entirely new method of investigation.

It is not possible to perform X-ray astronomy from the ground because the atmosphere is completely opaque in that part of the electromagnetic spectrum (which is fortunate for life on the surface). The first step was therefore to place instruments on sounding rockets launched to high altitude in order to perform observations for several minutes before falling back to Earth.

The first detection of X-rays from a celestial source came in 1948, when a rocket from White Sands Missile Range in New Mexico carried an instrument built by the US Naval Research Laboratory (NRL). The emission was found to be coming from the Sun.

Although the Sun is an intense source of X-rays when viewed from Earth, if it were to be relocated to the distance of the nearest star, some 4 light years away, the flux of radiation would be negligible owing to the diminishment of the inverse square law. However, during a 'solar flare' the X-ray intensity can increase by a factor of 1,000. Although some stars undergo optical variations suggestive of flaring, and X-rays from such events might be detectable in the case of the nearest stars, such episodes are infrequent and brief.

A rocket launched in 1956 with a detector built by Herbert Friedman at the NRL produced results that could either be interpreted as indicating that X-rays were being generated in Earth's atmosphere or were coming from a celestial source other than the Sun.

The first definite evidence of X-rays from beyond the solar system came in 1962, from a rocket flight carrying an instrument developed by the American Science and Engineering group headed by Riccardo Giacconi. Their plan was to find out if the Moon was fluorescing at X-ray wavelengths as a consequence of being irradiated by energetic solar wind particles, to obtain insight into the elemental composition of its surface[1], but the rocket veered off course and the instrument scanned another part of the sky, finding a strong source. The 100° field of view of the detector wasn't able to indicate the position of the source. In 1963 Friedman used a detector with a 10° field of view to localise it to the constellation of Scorpius, so it was designated Sco X-1. This rocket detected a weaker source that was subsequently identified by how it was occulted by the Moon; it proved to be the Crab Nebula in Taurus, the expanding shell of gas left by a supernova in AD 1054 that was so bright it could be seen in daylight.

When a rocket carrying a Massachusetts Institute of Technology instrument with a collimator to view a narrow field of view localised Sco X-1 to within 1° in 1964, astronomers were surprised to find no obvious optical counterpart. In

1. In fact, this was a ruse to obtain funding, because the government was eager to investigate the composition of the lunar surface and completely indifferent to proposals to determine whether celestial sources were emitting X-rays. The rocket's veering 'off course' was therefore no great disappointment to the scientists.

1966 the position was refined to within about 1/1,000[th] of a square degree of sky, which was sufficient for an optical search.

Attention soon settled on the variable star V818 Scorpii. The X-ray output is about 60,000 times the total luminosity of the Sun, and it displays a regular variation with a period of 18.9 hours. The optical counterpart also varies in an irregular manner on various timescales but there is no correlation with the X-rays.

When further optical work established Sco X-1 to be a member of a binary system, the initial supposition was that the primary star was transferring gas to a white dwarf companion and the X-ray emission originated from the infalling material. An alternative scenario became available with the discovery in 1967 of the first 'pulsar' by radio astronomers Jocelyn Bell and Anthony Hewish in Cambridge, which proved to be a rapidly spinning neutron star; the collapsed

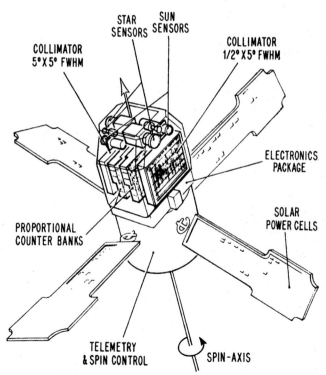

The Uhuru satellite, showing its axis of rotation and the side-viewing X-ray detectors. (NASA)

stellar core of a supernova. Sco X-1 is nowadays understood to be a neutron star that is just above Chandrasekhar's limit in orbit with a companion which has a mass substantially less than that of the Sun, although originally this star would have been the low-mass companion to the massive companion that underwent rapid evolution and became the neutron star.

To survey the entire X-ray sky, it was necessary to place instruments on a satellite. To do this, on 12 December 1970 NASA launched Explorer 42 as first of the three-spacecraft Small Astronomy Satellite series.

Because the launch took place from a platform off the coast of Kenya when that nation was celebrating the anniversary of its independence, the spacecraft was named Uhuru, which means 'freedom' in Swahili.

Uhuru had two X-ray telescopes installed back-to-back, and as the satellite slowly rotated, its instruments scanned the sky. Whenever a source of X-rays entered the field of view, the detector generated an electrical signal. This was transmitted to the ground. By knowing the orientation of the satellite in space, astronomers were able to estimate the position of the X-ray source. As soon as the sky had been surveyed, the satellite began again, to detect variations in the

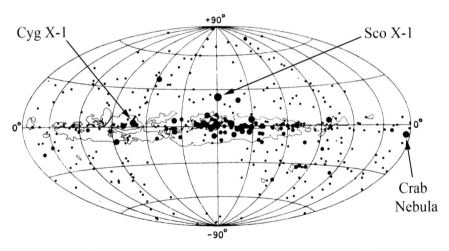

An all-sky map of the 339 X-ray sources detected by Uhuru. The plane of the Milky Way spans the horizontal axis. The indicated diameter of a source marker is proportional to the logarithm of its peak intensity. The three indicated sources are the Crab Nebula, Sco X-1 and Cyg X-1. (Ref: 'The fourth Uhuru catalogue of X-ray sources', W. Forman et al, *Astrophysical Journal Supplement*, vol. 38, pp. 357–412, 1978)

sources. It detected several hundred objects. Many were clearly outside of our galaxy, but a large proportion were so close to the plane of the Milky Way that they simply had to be 'local'.

An X-ray detector launched in 1964 from White Sands on a ballistic rocket had discovered a source in the constellation of Cygnus. Listed as Cygnus X-1, it was one of the strongest sources then known. Data between 1966 and 1968 indicated it to be a small source on the sky (in comparison to the Crab Nebula, at least) and to be variable. Uhuru revealed its intensity to vary on a timescale of a fraction of a second.

As a result of Albert Einstein's work, we know that nothing can travel faster than the speed of light, therefore an object cannot flicker faster than the time it would take light to span it. For example, an object that is 1 light second across cannot change its brightness faster than once per second. As the speed of light is 300,000 km/sec, this meant the source of the X-rays in Cygnus X-1 must be no larger than 10^5 km in diameter, which is only about one-tenth the diameter of the Sun.

In early 1971 the X-ray emission from Cygnus X-1 reduced in intensity and a radio source rose from below the threshold of detection. This was detected by Luc Braes and George K. Miley of Leiden Observatory, and independently by Robert M. Hjellming and Campbell Wade at the National Radio Astronomy Observatory in America.

The 1 arc-second accuracy of the radio position indicated the X-ray source was coincident with the 9th magnitude star listed in the Henry Draper catalogue as HDE 226868. A blue supergiant like that cannot emit appreciable amounts of X-rays. Astronomers therefore suspected the star must have a companion that was capable of heating gas to a temperature of millions of degrees to serve as the source of that radiation.

At the March 1971 meeting of the American Astronomical Society held in Baton Rouge, Louisiana, Giacconi boldly predicted Cygnus X-1 would prove to be a black hole. In reporting this, the *New York Times* not only capitalised the novel term, it also used inverted commas in the expectation that it would be new to members of the general public.

Confirmation that HDE 226868 was associated with the X-ray source came serendipitously in 1972 from an instrument on the Copernicus satellite, whose 'spiral scan' technique could localise a source to within about 10 arcsec.

The star field containing HDE 226868 (the bright star at the centre). (Palomar Observatory)

Working independently, Louise Webster and Paul G. Murdin at the Royal Greenwich Observatory in England, and Charles Thomas Bolton at the David Dunlap Observatory of the University of Toronto in Canada, reported in 1972 that their Doppler measurements of HDE 226868 implied the presence of a companion. There were no spectral lines from the secondary component, only those from the primary, which oscillated in wavelength with a period of 5.59983 days as the two components of the binary system revolved around the common centre of mass.

By presuming that we are viewing the binary system 'edge on' (i.e. with an inclination of 90°) and assuming a mass for the supergiant star, it is possible to

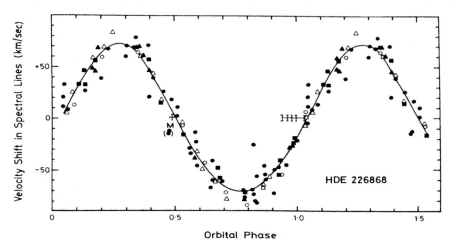

The Doppler shift in the spectrum of HDE 226868. (Cited in *X-ray Astronomy*, J. L. Culhane and P. W. Sanford, Faber and Faber, 1981)

calculate from the radial velocity of the visible star the *minimum* mass of the unseen object; its actual mass will then depend upon the inclination. The radial velocity curve establishes a well-defined determination of the 'mass function', a mathematical expression which involves the masses of the two stars and the inclination of the orbit.

The plane of the Cygnus X-1 system is sufficiently inclined to our line of sight to preclude eclipses by either component. The range of inclinations was estimated at 27–65°. For all sensible values of the mass of the supergiant and all plausible inclinations, the mass function required the unseen companion to exceed the maximum possible mass for a neutron star by a factor of about three. This required it to be a black hole.

Data from the first High Energy Astronomy Observatory revealed Cygnus X-1 to be 'flickering' on a timescale of milliseconds. This rapid variability in the X-ray flux meant the source could be no greater than 300 km in diameter, and only a black hole could compress 10 solar masses into a volume with that diameter.

So where do the X-rays come from? The Sun blows off material in the form of the solar wind, and larger, hotter stars have more intense stellar winds. The detailed profiles of the hydrogen-alpha lines in the spectrum of HDE 226868 indicate the star is surrounded by a gaseous envelope that is being accelerated

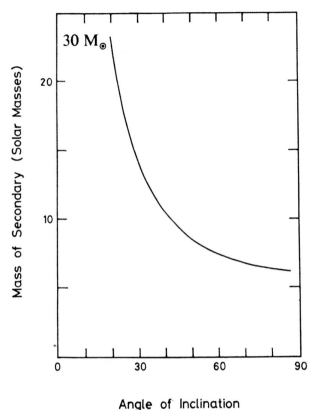

An analysis of the 'mass function' for HDE 226868 in which the primary is assumed to be 30 solar masses, indicating the secondary mass to exceed that which would allow it to be a neutron star at all non-eclipsing orbital planes, thereby establishing it to be a black hole. (Cited in *X-ray Astronomy*, J. L. Culhane and P. W. Sanford, Faber and Faber, 1981)

away at speeds of about 1,500 km/sec. Calculations suggest it sheds mass at a rate of 2.5×10^{-6} solar masses per year. Some of this outflow will pass within the gravitational influence of the black hole and be drawn into orbit around it.

This possibility had been envisaged in 1964 by Edwin Ernest Salpeter at Cornell when he realised that the angular momentum of gas attracted by the gravitation of a singularity would cause the material to spiral inward and form an 'accretion disk' around the black hole.

Of course, while the overall gravitational attraction of a black hole will be no greater than that of a star with the same mass at any given distance, in the case of a black hole material can approach much closer to the centre, hence the gravitational force in the *immediate vicinity* of the hole can be enormous.

By a well-established law, the inner portions of this accretion disk must be rotating more rapidly than the outer parts. The differential in speed will cause intense friction within the disk. Salpeter had predicted that as the temperature increases to millions of degrees, the material, by then a plasma of electrically charged particles, will emit X-rays. This had been independently predicted in 1965 by Zel'dovich, who pointed out that if one star in a binary system were to collapse to produce a black hole, the X-rays from the gas it was drawing off its companion ought to be detectable.

Thus Uhuru was detecting the superheated material in the inner part of the accretion disk, radiating X-rays as it spiralled in towards the central black hole. Flickering on a timescale of milliseconds is consistent with either turbulence in an accretion disk around a black hole or clumps of 'hot' gas orbiting extremely close to the event horizon. If HDE 226868 is 30 solar masses and the black hole is 15 solar masses, the radius of the event horizon will be a mere 44 km.

The X-ray flux varies with the phase of the orbit. It is least during superior conjunction when the orbiting objects are most closely aligned with the Earth and the compact source is the more distant. This means that the emissions are being partially blocked by circumstellar matter, very likely the stellar wind of HDE 226868 or perhaps the matter stream across to the accretion disk. Based on a model of stellar evolution, the primary star must be about 350,000 times as luminous as the Sun. This implies the compact object must be orbiting at a distance of only twice the radius of its primary. In such a situation, the surface of the primary must be tidally distorted into a tear-drop shape by the gravity of the massive companion, with the tip pointing toward the companion. Its shape will be further distorted by rotation. As a result, the optical brightness of the primary varies by 0.06 magnitudes during each 5.6 day binary orbit, with the minimum magnitude occurring when the system is most closely aligned to the line of sight.

As accreted matter falls inward, it loses gravitational potential energy. Part of this released energy is dissipated by a pair of 'jets' aligned perpendicular to the accretion disk, along which material is transported outward at a substantial fraction of the speed of light (i.e. relativistic velocities). The angle between the jets and the line of sight has been estimated at $30°$ and there is a periodicity of about 300 days in the emission which may be caused by the precession of the accretion disk.

The plasma that spirals around the accretion disk will 'wind up' an intense magnetic field with its axis centred on the black hole and perpendicular to the disk. This will produce a pair of columnar tubes that accelerate plasma away in either direction.

Although the material travelling along the Cygnus X-1 jets does not emit in the optical spectrum, one of them interacts with a relatively dense part of the interstellar medium, making an energised ring detectable at radio wavelengths. This also appears to be creating an optical nebula. There is no corresponding ring in the other direction, since that jet is facing a lower density portion of the interstellar medium.

Since HDE 226868 shares a common motion through space with a stellar association known as Cygnus OB3 located about 6,000 light years away, this suggests that Cygnus X-1 is about 5 million years of age. The largest stars in OB3 are around 40 solar masses. As more massive stars evolve more rapidly, this implies the progenitor star for Cygnus X-1 must have had at least 40 solar masses. That star would have shed a lot of material with a stellar wind. When and how the black hole formed is not known. It could be the relic left behind when the star suffered a supernova explosion, in which case the ejected shell

An artist's explanation of the HDE 226868 system, showing the matter stream from the primary to the accretion disk which, like the pair of relativistic jets, is centred on the (invisible) black hole. (NASA/CXC/M.Weiss)

has dispersed, or some other circumstance might have prompted it to collapse directly into a black hole.

In December 1974 Cygnus X-1 became the subject of a friendly wager by physicists Stephen Hawking and Kip Thorne. This took the form of a single handwritten sheet in which Thorne asserted that the system contained a black hole and Hawking insisted it didn't. At stake was a year's subscription for the American racy magazine *Penthouse* to Thorne or four years' of the British satirical *Private Eye* to Hawking. In June 1990, Hawking conceded and pornography started to arrive in Thorne's mail, much to the dismay of his wife.

Meanwhile, Hawking had made a momentous discovery about black holes. In 1973 Jacob Bekenstein had claimed that the area of an event horizon was a direct measure of the entropy of the black hole that it hosted. In attempting to prove this assertion incorrect, and finding it to be true, Hawking realised that if quantum mechanics and relativity were both taken into account in describing an event horizon, it was possible for a black hole to lose energy (by a process that became known as Hawking radiation) and ultimately evaporate.

Although Cygnus X-1 was the first stellar-mass black hole to be identified, others were soon located. Furthermore, black holes with masses ranging from millions to billions of solar masses have been found in the cores of galaxies as the agents responsible for quasar activity.[2]

The similarities between the emissions of X-ray binaries like the one which contains Cygnus X-1 and active galactic nuclei suggests a common mechanism of energy generation involving a black hole, an orbiting accretion disk, and a pair of jets. For this reason, Cygnus X-1 is one of a class of objects known as micro-quasars.[3]

As a final thought, because the only properties intrinsic to a black hole are its mass, spin and charge, it bears a striking similarity (on a gigantic scale) to a fundamental subatomic particle, and perhaps that is precisely what a black hole is in some grander scheme of things than we can perceive.

I am grateful to Ronald W. Hilditch of my *alma mater*, the University of St Andrews in Scotland, for making several suggestions that improved my text.

---

2. I discussed super-massive black holes in the Yearbook of Astronomy 2018.

3. I discussed SS433, the archetypal micro-quasar, in the Yearbook of Astronomy 2019.

'Oumuamua – Interstellar Interloper

Neil Norman

The sky survey team of Pan-STARRS (The Panoramic Survey Telescope and Rapid Response System) which is located at the Haleakalā Observatory, Hawaii are constantly finding new objects in the sky. To date, they have 180 comets and thousands of asteroid discoveries to their name, but on the night of 19 October 2017 they discovered an object which was to rewrite the science books.

Astronomer Robert J. Weryk was collecting data during his shift and noted a very fast moving object travelling through the constellation of Cetus.

Resembling a splinter fragment from a once larger body, no other object in the Sun's family is known to show characteristics like those shown in this artist's impression of 'Oumuamua, the first known interstellar object to pass through our Solar System. (ESO/M. Kornmesser/ Wikimedia Commons)

Fast moving objects mean only one thing – it was close to Earth. At the time of discovery, 'Oumuamua was located at a distance of around 0.22 AU (33,000,000 km) from Earth and already heading away from the Sun.

Weryk submitted the astrometry to the Minor Planet Center who used the data to determine a preliminary orbit. This turned out to strongly resemble that of a comet, hence its initial designation of C/2017 U1. However, with no cometary activity being observed, the object was re-classified as an asteroid and given the new designation of A/2017 U1 (the A identifying it an asteroid). Further refinements to the orbit led astronomers to the conclusion that this object was in fact an interstellar interloper, having arrived indeed from beyond our Solar System.

A new designation 1I/2017 U1 was assigned by the International Astronomical Union (IAU), the prefix '1I' referring to the fact that this was the first Interstellar object. We now identify the object as 1I/2017 U1 and with the Hawaiian name 'Oumuamua to honour the place from where it was discovered. A rough translation of the name is "scout or a messenger from our distant past reaching out to us from far away".

The first suggestion for a name was actually "Rama", this in a nod to the 1973 science fiction novel *Rendezvous with Rama* by the English writer Arthur C. Clarke which tells of an alien spacecraft discovered under similar circumstances.

Observations

Discovered 40 days after it made its closest approach to the Sun, the object had already passed through perihelion on 9 September, but had not been recorded in any STEREO HI-1A observations (STEREO being the NASA funded Solar Terrestrial Relations Observatory, consisting of two virtually identical spacecraft launched in October 2006 to study the Sun). This suggested that the object had a magnitude of only 13.5 or lower.

At the time of discovery 'Oumuamua was at magnitude 19 even though it was only 0.21 AUs from Earth. By the end of October the magnitude was 23 and by the end of the year it had dropped to magnitude 27, at which point it was already 2.8 AUs from Earth and too faint to be seen by even the largest telescopes on Earth.

The Allen Telescope Array at Hat Creek Observatory, California, USA and the SETI (Search for Extraterrestrial Intelligence) radio telescope examined

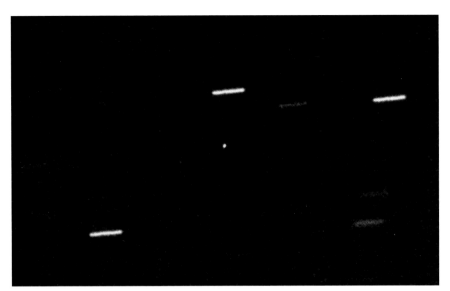

An image of 'Oumuamua taken on 28 October 2017 with the 4.2 metre William Herschel Telescope on La Palma in the Canary Islands. 'Oumuamua is visible as a stationary point of light at the centre of the image. The trails are those of background stars which appear streaked due to the fact that the telescope was tracking the rapidly moving asteroid at the time. (NASA/Alan Fitzsimmons/Isaac Newton Group/Wikimedia Commons)

'Oumuamua, although no unusual radio emissions were detected. This was followed by examination for narrowband signals with the Breakthrough Listen Hardware and the Green Bank Telescope, although again nothing was found.

From Whence It Came?

Not long after discovery astronomers came to the conclusion that 'Oumuamua was the first unbound object detected travelling through the Solar System. This posed the question as to where it originally came from. The typical orbit of an unbound object is hyperbolic, with the object being drawn towards the Sun but eventually escaping from the source of gravity and back into interstellar space.

'Oumuamua was found to have the greatest eccentricity value (1.20) recorded for any object ever seen travelling within the Solar System (any eccentricity value in excess of 1 indicating that the object in question exceeds the Suns' escape velocity) and is therefore not bound by the gravitational influence of our parent star. The previous record holder was a comet named C/1980 E1 (Bowell), although its eccentricity was boosted by a close pass of Jupiter as it approached the Sun. The difference with 'Oumuamua is that it has not gained any orbital energy since it entered our planetary realm.

'Oumuamua was seen to approach the Sun from roughly the direction of the Solar Apex, this being the point in space (located in the general direction of the constellation Lyra) towards which our Sun is heading. Its incoming motion was 6 degrees from the apex. The Catalina Sky Survey found two precovery images in their data from 14 and 17 of October respectively to help confirm the positioning.

The most conclusive information gained from all the data obtained, including an observed velocity of 26.33 km/s (58,900 mph) while it was still beyond the Solar System, was that 'Oumuamua was of interstellar origin. Indeed, a constant velocity this high can only mean it has come from beyond.

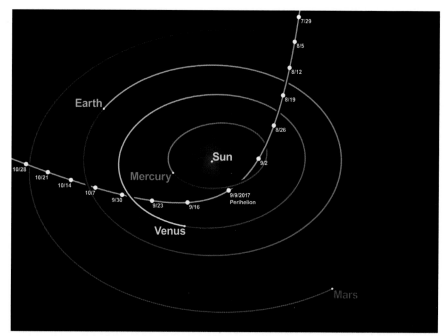

The passage of 'Oumuamua through the inner planets as it passed through perihelion in September 2017, travelling beneath the orbital planes of Mercury, Venus and Earth before crossing the ecliptic plane at around the orbital distance of Mars. (Tom Ruen/Wikimedia Commons)

'Oumuamua entered from a steep angle above the plane of the ecliptic. By the time of perihelion it was travelling at 87.71 km/s (196,200 mph) as it passed below the ecliptic on 6 September before making a sharp hairpin turn on 9 September at a distance of 0.25 AU (37,400,000 km / 23,200,000 miles) from the Sun (appreciably nearer to our parent star than Mercury at its closest approach). 'Oumuamua then rapidly sped north towards the constellation of Pegasus at an angle of 66 degrees from its direction of approach.

As it started to move away from the Sun, 'Oumuamua passed below the orbit of the Earth on 14 October at a relative close distance of 0.1616 AU (24,175,015 km / 15,021,658 miles), passing over the ecliptic plane on 16 October and over the orbit of Jupiter in May 2018. It was due to cross the orbit of Saturn in January 2019, and eventually pass beyond Neptune in 2022, still travelling in the general direction of Pegasus. By this time its velocity will have decreased to a value roughly equal to the velocity it had before approaching the Sun.

Origin

As already mentioned, 'Oumuamua was seen to approach the Sun from roughly the direction of the Solar Apex, this being the direction from which these objects would be seen to come as we meet them "head-on", so to speak. However, this does not necessarily mean that it originated from the star Vega.

A mean motion of 26.33 km/s (which is the velocity close to the mean motion of material within the Milky Way closest to the Sun and known as the local standard of rest) means that 'Oumuamua could easily have originated from a far more distant place within the galaxy than Vega. Even if we locate any particular star that 'Oumuamua has appeared to come from, we must be cautious in that it may have simply been passing by that star as it did our Sun. In millions of years from now an alien civilisation may be able to track it back to our star, although that would not mean that it had originated here.

A figure for exactly how long it has been passing through our Galaxy can not be given precisely and our Sun could easily be the first star it has encountered since its ejection from its place of origin. We do know that around 1.3 million years ago 'Oumuamua passed within 0.52 light years of the nearby star TYC 4742-1027-1, although its velocity is too high for it to have originated there, and it probably passed through that star system's Oort cloud at a speed of 103 km/s (230,000 mph).

Comet or Asteroid?

Although originally classed as a comet, scientists are still trying to determine what exactly 'Oumuamua is. Comets were the first objects expected to be seen entering the Solar System, yet 'Oumuamua certainly did not act like a comet in that no coma was detected (even with deep images) and no tail was seen ever. But does that necessarily mean it can not be a comet?

Not at all! 'Oumuamua could easily have built up a crust of irradiated organic compounds (tholins) several feet thick as it traversed through space, which would mean that the volatile materials within could not escape to create a coma or tail.

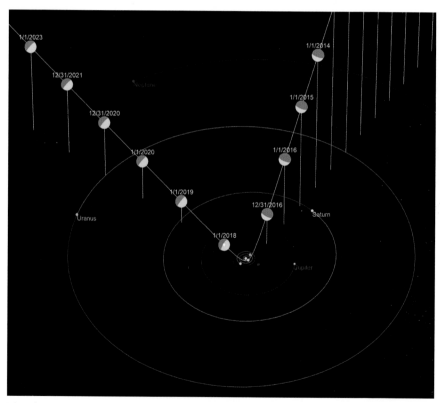

This diagram depicts the steeply inclined trajectory of 'Oumuamua through the inner Solar System showing how it plunged towards perihelion within the orbit of Mercury in September 2017, before receding at a steep angle back towards interstellar space. (Tom Ruen/Wikimedia Commons)

In June 2018, scientists revealed that 'Oumuamua had gained additional speed from somewhere, which may suggest that at some point a jet had formed creating a rocket thrust scenario, again strengthening the comet theory.

Composition and Characteristics

Light curves obtained indicated that it was a rather odd elongated shape, far more elongated than anything seen within our Solar System to date. Although no direct visual observation was ever made, the longest-to-shortest axis ratio could be 5:1 or larger. If we assume an albedo of 10 percent and a 6:1 ratio, 'Oumuamua has dimensions of ~230 metres x 35 metres x 35 metres with an average diameter of ~110 metres.

'Oumuamua also travels through space by rotating around a non-principal axis, resulting in a tumbling motion. This accounts for the different periods of rotation given such as 7.3 hours and 8.1 hours. It is very likely that 'Oumuamua underwent a collision within its star system before being ejected and has been tumbling ever since.

The spectrum is very similar to that of D-type asteroids and it is red in colour. The composition also suggests that it is comprised of dense metal-rich rock, but detailed analysis cannot be clarified because of the tholins that encase the pristine material beneath.

Astronomers calculate that several objects like this must pass by Earth each year, and that there is a potential of several thousand within the orbit of Neptune on any given day. With this quantity of objects, one must wonder why only one has been seen. The reasons for this include the fact that the velocity of these objects is so high. In addition, their small size means that they will have a very low magnitude even when close, typically magnitude 23 at the distance of 1AU from the Sun. It is literally a question of having your telescope pointing in the right direction at the right time. This is a task easier said than done.

Forgotten on the Moon

Bill Leatherbarrow

With the passing of the British-born astronomer Ewen Whitaker in 2016, at the grand old age of 94, lunar science lost one of its greatest practitioners of the twentieth century. From relatively modest beginnings as an amateur lunar observer and Director of the British Astronomical Association's Lunar Section, Whitaker moved to the USA in 1958 to join Gerard Kuiper's team of professional lunar mappers. Over the following years he did much to prepare the way for the successful Apollo Moon landings of the late 1960s and early 1970s.

But apart from his pioneering work as a lunar cartographer, Whitaker also wrote the standard textbook on the history of how the lunar surface had been mapped since the time of Galileo and how its features had acquired the names they bear today. *Mapping and Naming the Moon* appeared in 1999. Still valuable today, it charts the circuitous progress of lunar nomenclature from the earliest efforts of Michel van Langren (better known as Langrenus), whose map of 1645 allocated names of political and scientific luminaries to the major topographic features and was thus based upon patronage and homage. Whitaker's narrative follows the attempts of later cartographers like Giovanni Riccioli to rationalise the system of lunar nomenclature, and finally describes the current official system, developed under the auspices of the International Astronomical Union in the course of the twentieth century. In the accepted IAU system features on the lunar surface are now usually named after deceased scientists or explorers who have made outstanding contributions to their field. Proposed names should not duplicate any existing lunar name.

There is no doubt that, in due time, the name of Ewen Whitaker will be attached to a lunar feature, and nothing could be more fitting. However, the history of lunar nomenclature is littered with the names of those who fell by the wayside and whose contributions, large and small, have been denied recognition on the Moon by the vagaries of an evolving nomenclature system. Considerations of space do not allow us to identify every instance here, for there are very many, but the following examples of some of those who have been forgotten on the Moon should serve as an acknowledgement of achievements that have been overlooked or excluded, but are nevertheless worth remembering.

The crater Sheepshanks located on the Mare Frigoris is named after Anne Sheepshanks (1789–1876), sister of the English astronomer Richard (1794–1855) in whose name she left a legacy to Cambridge University, endowed a scholarship and donated her brother's books to the Royal Astronomical Society following his death. Anne's role as a benefactor has been amply rewarded, but Richard's astronomical work remains unacknowledged on the Moon. Similarly, the great American astronomer Harlow Shapley (1885–1972) is rightly commemorated on the Moon by the crater Shapley on the southern boundary of the Mare Crisium. However, the contribution made by his wife Martha Betz Shapley (1890–1981), who provided invaluable assistance to Shapley's researches at Mount Wilson and Harvard, is unmarked. In the same way, the work of the German astronomer and comet discoverer Gottfried Kirch (1639–1710) is marked by a small 11km

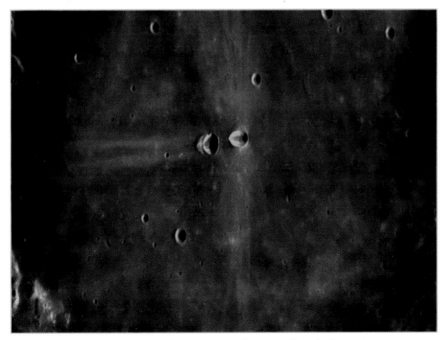

Messier (right) and Messier A, once known as W. Pickering. (Bill Leatherbarrow)

crater north of Archimedes on the Mare Imbrium, whereas the contribution made by his second wife Maria Margaretha Kirch (née Winckelmann) (1670–1720), who assisted him in many of his observations, is not represented. It might be argued that Maria's astronomical work was comparable in importance to that of her husband, but she never achieved the same level of formal recognition from the scientific community – something she put down to gender bias.

No such bias accounts for how the Clark family of telescope makers are represented on the Moon. Alvan Clark & Sons are justly recognised as the makers of the optics of some of America's finest and largest refractors, including the 36-inch Lick refractor and the 40-inch Yerkes telescope, the latter still the world's largest working refractor. It is no surprise therefore that father Alvan (1804–1887) and younger son Alvan Graham (1832–1897) are commemorated in the 27km crater Clark, situated on the far side of the Moon. What is perhaps unusual is that the elder son George Bassett Clark (1827–1891) is not included in this family plot. It could be argued that this situation is not so much a slight on George as recognition of Alvan Graham Clark's achievement in making the first telescopic observation of the white dwarf companion to the star Sirius in 1862.

There are many other instances of one member of a partnership going unrecognised in favour of the other, but a particularly interesting example is that of the Pickering brothers. Edward Charles Pickering (1846–1919) was director of Harvard Observatory and he did much to help found the discipline of astrophysics; his younger brother William Henry (1858–1938) was an outstanding planetary observer whose contributions were marred by an extravagant imagination when it came to explaining the nature of what he observed. The lunar cartographer Johann Nepomuk Krieger (1865–1902) and his collaborator Rudolf König (1865–1927) named separate craters after both Pickering brothers in Krieger's *Mond Atlas* (1912). However, in 1967, the IAU deleted W. H. Pickering's name from the small crater on the Mare Fecunditatis to which it had been attached, and that crater reverted to its previous label of Messier A. Instead William Henry had to move in with his elder brother and the crater once known as E. Pickering, located near the centre of the lunar disc as seen from Earth, now commemorates both brothers. Interestingly, both Pickering brothers are also commemorated by a single crater on Mars.

Until the IAU stamped its authority on lunar nomenclature, the naming of lunar features had been haphazard and largely left in the hands of the authors of lunar maps. In 1913, under the auspices of the Lunar Nomenclature Committee of the International Association of Academies, Mary Adela Blagg and Samuel Arthur Saunder had attempted to bring some semblance of order by publishing a collated list of named formations in the maps of Neison, Schmidt and Mädler. Until the middle of the twentieth century most lunar mappers had been amateurs, and to a large extent they followed the nomenclature used on earlier maps as regularised by Blagg and Saunder. However, they tended to introduce unregulated new names for features that had previously gone unnamed or, following a system introduced by Mädler, had been designated by the name of a nearby named feature followed by a simple letter (for example, the renaming of Messier A as W. Pickering). The great British selenographer Walter Goodacre (1856–1938) generally followed Neison in naming formations on his map of 1910, with the exception that he allocated letters to a few previously unnamed structures. However, Hugh Percy Wilkins (1896–1960), Goodacre's younger colleague and an eventual successor as Director of the Lunar Section of the British Astronomical Association, adopted a far more radical approach to nomenclature in producing a succession of maps during the first half of the twentieth century.

By the time his 300-inch map appeared in 1946 (originally drawn to a diameter of 300 inches, but printed at a reduced scale of 100 inches), Wilkins had proposed a raft of about 100 new names, most of them recognising his friends and colleagues both in the BAA Lunar Section and abroad. Many of these were amateur observers who had spent their lives teasing out the fine lunar details that Wilkins included on his over-cluttered maps. They included

Ewen Whitaker (left) and Hugh Percy Wilkins at Greenwich, circa 1953. (Walter Haas / BAA Lunar Section Archive)

Ewen Whitaker, who succeeded Wilkins as Lunar Section Director, as well as skilled observers of the time such as Keith Abineri, Leslie F. Ball, Robert Barker, Richard Baum, Walter Haas, Harold Hill and Patrick Moore, to name but a few. There is no doubt that people such as these deserve to be remembered in some way for their painstaking work at a time when professionals took little or no interest in the Moon, but the fact is that Wilkins' proposals did not meet the criteria of the IAU and they were rejected at the 1948, 1952 and 1955 meetings of the General Assembly of that body. Some of Wilkins' names (e.g. those of the polar explorers Amundsen and Scott) were restored in the *Rectified Lunar Atlas* produced by D. W. G. Arthur and Ewen Whitaker in 1963 and approved by the IAU the following year. But many amateur observers who had taken pride in the recognition briefly afforded by Wilkins found themselves once again forgotten on the Moon.

Among such people, a couple stand out. Harold Hill (1920–2005), a draughtsman in his professional life, spent more than half a century in the painstaking telescopic exploration of the lunar surface, producing hundreds of drawings that combined rigorous topographical accuracy with exquisite artistry. He was in my view the finest lunar observer of the twentieth century. His book *A Portfolio of Lunar Drawings*, published in 1991, not only displays his remarkable gifts to perfection, but has also served as an inspiration for subsequent generations of observers. However, Hill was not a professional scientist, and indeed his work is of greater aesthetic than scientific significance. His determined attempts to capture the lunar surface under all angles of illumination and to chart the difficult regions of the lunar south pole were eventually rendered redundant by spacecraft imagery. As a result he does not meet the IAU criteria, even though the impact of his vision still resonates in the world of lunar study.

Several of the great British amateur selenographers, such as Neison, Elger, Goodacre and Wilkins, have been honoured with named craters on the Moon. Most of us would expect the same to be true of Patrick Moore (1923–2012), perhaps the best-known amateur astronomer and lunar observer of all time. Famous for his scores of popular books and his long-running television programme *The Sky at Night*, Moore was responsible for bringing lunar science and astronomy in general to the widest of audiences. Most astronomers alive today, both amateur and professional, would readily acknowledge his influence

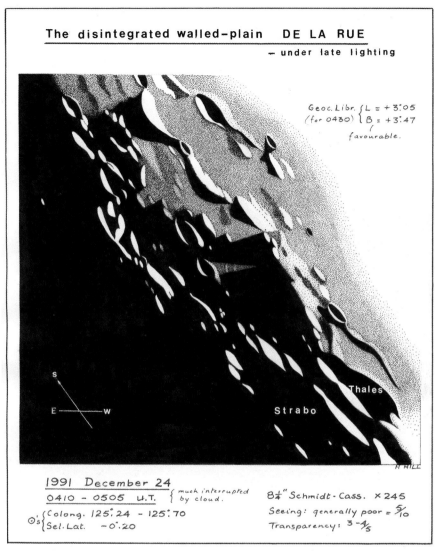

The disintegrated walled-plain **DE LA RUE**

– under late lighting

Geoc. Libr. { L = +3°.05
(for 0430) { B = +3°.47
↑
favourable.

S
E —————— W

Thales

Strabo

H HILL

1991 December 24

0410 – 0505 U.T.   { *much interrupted by cloud.*

⊙'s { *Colong. 125°.24 – 125°.70*
     { *Sel. Lat.   – 0°.20*

8¼" *Schmidt-Cass. × 245*
Seeing: generally poor = 5/10
Transparency: 3 – 4/5

Drawing of crater De la Rue by Harold Hill. (BAA Lunar Section Archive)

on their choice of interest or career. Yet, as an amateur like Hill and despite being an honorary Fellow of the Royal Society and a member of the International Astronomical Union, Patrick Moore fails to meet current IAU nomenclature criteria and is thus forgotten on the Moon, if not by those who study it. There is a lunar crater Moore, located on the far side, but it is named after the American astronomer Joseph Haines Moore.

The treatment of figures like Moore and Hill might seem unfair and counterintuitive, but it is by no means perverse. It has taken a long time to secure international agreement for an ordered system of nomenclature, and few would wish to return to the unregulated chaos of the past. However, as with everything in life, there is a price to be paid for this order, and unfortunately it has been paid by those who are now 'Forgotten on the Moon'.

Miscellaneous

Some Interesting Variable Stars

Roger Pickard

The night sky is full of many types of variable star, some of which change in brightness by a significant amount over a period of a few days or even hours, others doing so more slowly, taking weeks or even months to show much change.

So, if there are no planets around to observe that night, or you are tired of looking at faint and fuzzy objects, why not try your hand at a variable star or two, especially if the Moon is out of the way? In the following lists, some stars will be found easier than others to locate and estimate, but none should be too difficult. In addition, some charts show more than one variable star and you are most welcome to have a go at any of them besides the ones I have highlighted here. Do also note that some stars, particularly the telescopic ones, will have more than one chart to help you identify the field of view. Comparison charts for ALL of these stars can be found on the British Astronomical Association Variable Star Section website at **www.britastro.org/vss**

In the tables, each star is followed by its type (for example, SRb indicates that it is a semi-regular star of type b – for further explanation of type see the American Association of Variable Star Observers Variable Star Index at **www.aavso.org/vsx**) and then its position. Next we have the typical range in magnitude of the star, then its period (where applicable) and the chart it can be found on, the chart reference referring to the number that can be found at the top left of each chart and which uniquely identifies each one. For example, on the ST Cam chart the reference is given as 111.02 which can be seen at the top left of the chart. Finally, a recommendation of how frequently the star should be observed is given. Below each table are some additional notes that I feel may be helpful. Comparison charts for all the stars mentioned can be found at **http://www.britastro.org/vss/xchartcat/index.html** Light curves for many of the stars listed in the following tables are located at the end of the article. Note that the light curves are shown with a starting date of 1 January 2015 (with the exception of AH Dra which starts on 1 July 2015).

Although I have listed only five stars under *Binocular Variables*, several of the stars listed as *Telescopic Variables* can be followed with binoculars when bright. Indeed, a few of them can only be followed in binoculars when at or near their brightest.

Binocular Variables

STAR	TYPE	RA		DECLINATION		RANGE	PERIOD (D)	CHART	FREQUEN...
		H	M	°	′				
ST Cam	SRb	04	51	+68	10	6.3 / 8.5	300	111.02	7 days
AH Dra	SRb	16	48	+57	49	6.4 / 8.6	158	106.03	7 days
RX Lep	SRb	05	11	−11	51	5.1 / 6.7	80	110.02	7 days
X Oph	M	18	38	+08	50	5.9 / 8.6	338	099.02	7 days
Z UMa	SRb	11	57	+57	52	6.2 / 9.4	195.5	217.02	7 days

- **ST Cam** had been showing fairly regular variations between about magnitude 6.8 and 8+, but since late 2017 these have faded to around 7.5 and 8+. It will be interesting to see how it has performed since then, and how it continues to perform.
- **AH Dra** has a reasonable regular variation between about magnitudes 7 and 8.5.
- **RX Lep** has a range of about 2 magnitudes and there is a hint that there could also be a much longer period of around 20 years.
- **X Oph** shows a lovely regular variation between about magnitudes 6.8 and 8.6 (note – not quite as bright as the range shown in the table). There is also an 8.7 magnitude companion 0.5″ away.
- **Z UMa** is an easy star to find, being located little more than 3 degrees from Megrez (Delta Ursae Majoris), one of the stars forming the "bowl" of the "Plough".

Telescopic Variables

STAR	TYPE	RA		DECLINATION		RANGE	PERIOD (D)	CHART	FREQUENCY
		H	M	°	′				
R And	M	00	24	+38	35	5.8 / 15.2	409.2	053.02	7 days
V Boo	SRa	14	30	+38	52	7.0 / 12.0	258.01	037.02	7 days
CI Cyg	EA/G + Z And	19	50	+35	41	9.1 / 11.7	855	006.02	Nightly
U Gem	UGSS + E	07	55	+22	00	8.2 / 14.9	105	008.04	Nightly
R Hya	M	13	30	−23	17	3.5 / 10.9	380	049.03	7 days
U Ori	M	05	55	+20	11	4.8 / 13.0	377	059.02	7 days
AX Per	Z And + E	01	36	+54	16	9.5 / 12.8V	680.83	073.02	Nightly

- **R And** is one of the first variables I followed, although I could rarely see it at minimum as it can reach below magnitude 15. However, it was always interesting to see how quickly it recovered and I could follow it again.
- **V Boo** showed quite a large range of around 4 magnitudes back in the 1920s and 1930s, before dropping to around 1.5 magnitudes between about 1990 and 2010. However, this has been slowly increasing over the last few years and is now some 2.5 magnitudes, varying between almost 7 and 9.6.
- **CI Cyg** is a "symbiotic" variable of the Z And class, by which we mean it is a close binary consisting of a hot star and a star of late type with an extended envelope excited by the radiation from the hot star. It also shows eclipses. There is a lot of scatter in the light curve, so you need to take care with this one.
- **U Gem** is the prototype of its class and although it spends most of its life below about magnitude 14, it also shows regular outbursts every 100 days or so, taking it up to magnitude 9 before fading almost immediately to its minimum brightness.
- **R Hya** is one of the earliest known variable stars. Its period has undergone a long term reduction since its variability was first discovered during the late 17th century.

- **U Ori** is a beautiful, fairly regular long period variable (LPV, or M for Mira type) becoming quite bright at maximum. However, being in Orion, you can not follow it throughout the whole year, hence the gaps in the light curve.
- **AX Per** is similar in many ways to CI Cyg. However, around the year 2000 it showed much smaller variations of less than about 1 magnitude. These have now grown to over 2 magnitudes and it will be interesting to see if the range shown in and around/during the 1990s of some 3.5 magnitudes can once again be attained.

Eclipsing Binary Stars

STAR	TYPE	MAX	MIN II	MIN I	PERIOD (days)	ECLIPSE DURATION (hours)
TV Cas	EA/SD	7.2	7.3	8.2	1.81	8
U Cep	EA	6.8	6.9	9.4	2.49	9
β Per (Algol)	EA	2.1	2.2	3.4	2.87	9.6

- Both **TV Cas** and **U Cep** would benefit from more intense coverage of their eclipses.
- As one of the brightest eclipsing binary stars in the sky, **β Per (Algol)** is well worth looking at, especially as it is so easy to find.

More information on this type of variable star can be found in the article *Eclipsing Binaries* by Tracie Heywood elsewhere in this volume.

The comparison chart for Algol is included at the end of this article, along with predictions for timings of minimum brightness for the star throughout 2020.

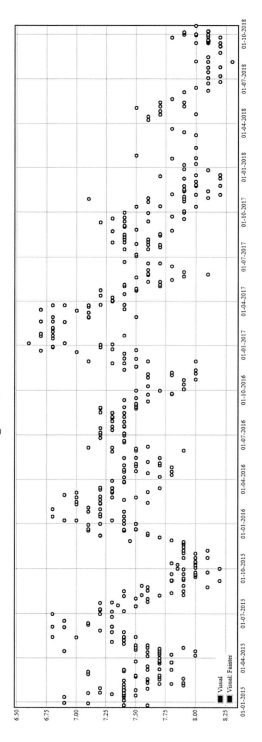

Light Curve for ST CAM

Light Curve for AH DRA

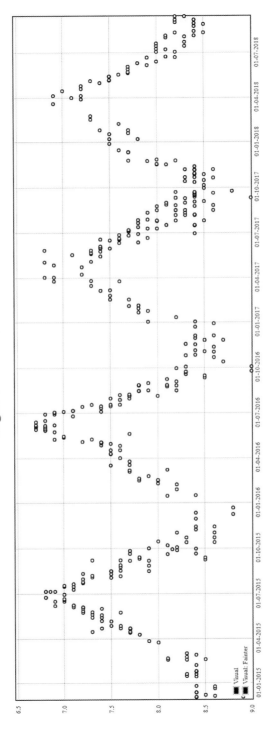

Light Curve for X OPH

Light Curve for Z UMA

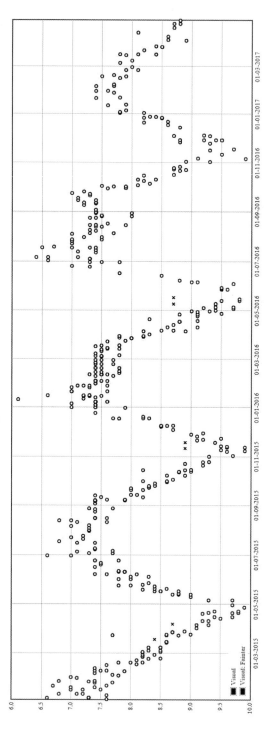

Light Curve for R AND

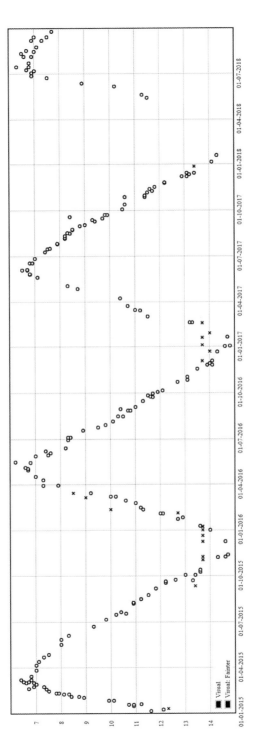

Light Curve for V BOO

Light Curve for CI CYG

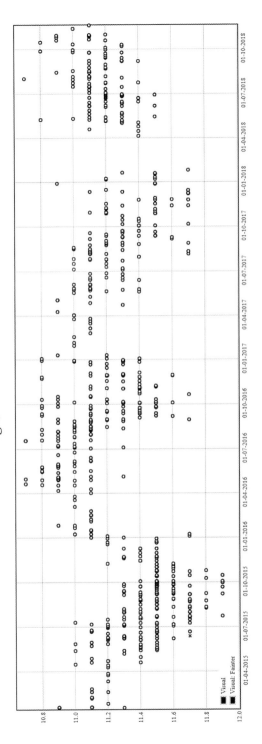

Light Curve for U ORI

Light Curve for AX PER

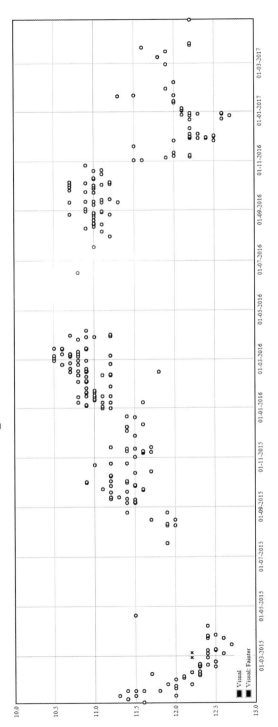

BETA PERSEI   03h 08m 10·1s +40° 57´ 20″ (2000)

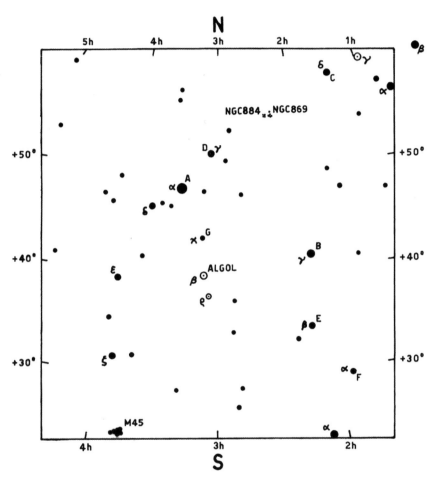

CHART:	A 1·8	E 3·0	BAA VSS
NORTONS STAR ATLAS	B 2·1	F 3·4	EPOCH: 2000
	C 2·7	G 3·8	DRAWN: JT 19-06-11
SEQUENCE:	D 2·9		APPROVED: RDP
HIPPARCOS VJ			

Minima of Algol in 2020

Beta (β) Persei (Algol): Magnitude 2.1 to 3.4 / Duration 9.6 hours

Month	Day	h		Month	Day	h		Month	Day	h		Month	Day	h	
Jan	2	19.5		Feb	3	8.5		Mar	3	0.6	*	Apr	3	13.6	
	5	16.3			6	5.3			5	21.4	*		6	10.4	
	8	13.1			9	2.1	*		8	18.3			9	7.2	
	11	10.0			11	22.9	*		11	15.1			12	4.0	
	14	6.8			14	19.7			14	11.9			15	0.9	
	17	3.8	*		17	16.6			17	8.7			17	21.7	
	20	0.4	*		20	13.4			20	5.5			20	18.5	
	22	21.2	*		23	10.2			23	2.3	*		23	15.3	
	25	18.0			26	7.0			25	23.2	*		26	12.1	
	28	14.8			29	3.8	*		28	20.0			29	8.9	
	31	11.7							31	16.8					
May	2	5.8		Jun	2	18.7		Jul	1	10.9		Aug	1	23.9	
	5	2.6			5	15.5			4	7.7			4	20.7	
	7	23.4			8	12.4			7	4.5			7	17.5	
	10	20.2			11	9.2			10	1.3			10	14.3	
	13	17.9			14	6.0			12	22.1			13	11.1	
	16	13.8			17	2.8			15	19.0			16	7.9	
	19	10.7			19	23.6			18	15.8			19	4.7	
	22	7.5			22	20.4			21	12.6			22	1.6	*
	25	4.3			25	17.3			24	9.4			24	22.4	
	28	1.1			28	14.1			27	6.2			27	19.2	
	30	21.9							30	3.0			30	16.0	
Sep	2	12.8		Oct	1	5.0		Nov	1	18.0		Dec	3	6.9	
	5	9.6			4	1.8	*		4	14.8			6	3.7	*
	8	6.5			6	22.6	*		7	11.6			9	0.5	*
	11	3.3			9	19.4			10	8.4			11	21.4	*
	14	0.1	*		12	16.2			13	5.2			14	18.2	
	16	20.9			15	13.1			16	2.0	*		17	15.0	
	19	17.7			18	9.9			18	22.8	*		20	11.8	
	22	14.5			21	6.7			21	19.7			23	8.6	
	25	11.4			24	3.5	*		24	16.5			26	5.5	
	28	8.2			27	0.3	*		27	13.3			29	2.3	*
					29	21.1			30	10.1			31	23.1	*

Minima marked with an asterisk (*) are favourable from the British Isles, taking into account the altitude of the variable and the distance of the Sun below the horizon (based on longitude 0° and latitude 52°N)

Note that during the winter months some favourable partial eclipses may also be visible.

Some Interesting Double Stars

Brian Jones

The accompanying table describes the visual appearances of a selection of double stars. These may be optical doubles (which consist of two stars which happen to lie more or less in the same line of sight as seen from Earth and which therefore only appear to lie close to each other) or binary systems (which are made up of two stars which are gravitationally linked and which orbit their common centre of mass).

Other than the location on the celestial sphere and the magnitudes of the individual components, the list gives two other values for each of the double stars listed – the angular separation and position angle (PA). Further details of what these terms mean can be found in the article *Double and Multiple Stars* published in the 2018 edition of the Yearbook of Astronomy.

Double-star observing can be a very rewarding process, and even a small telescope will show most, if not all, the best doubles in the sky. You can enjoy looking at double stars simply for their beauty, such as Albireo (β Cygni) or Almach (γ Andromedae), although there is a challenge to be had in splitting very difficult (close) double stars, such as the demanding Sirius (α Canis Majoris) or the individual pairs forming the Epsilon (ε) Lyrae 'Double-Double' star system.

The accompanying list is a compilation of some of the prettiest double (and multiple) stars scattered across both the Northern and Southern heavens. Once you have managed to track these down, many others are out there awaiting your attention …

Star	RA h	m	Declination °	'	Magnitudes	Separation (arcsec)	PA °	Comments
Beta[1,2] (β[1,2]) Tucanae	00	31.5	−62	58	4.36 / 4.53	27.1	169	Both stars again double, but difficult
Achird (η Cassiopeiae)	00	49.1	+57	49	3.44 / 7.51	13.4	324	Easy double
Mesarthim (γ Arietis)	01	53.5	+19	18	4.58 / 4.64	7.6	1	Easy pair of white stars
Almach (γ Andromedae)	02	03.9	+42	20	2.26 / 4.84	9.6	63	Yellow and blue-green components
32 Eridani	03	54.3	−02	57	4.8 / 6.1	6.9	348	Yellowish and bluish
Alnitak (ζ Orionis)	05	40.7	−01	57	2.0 / 4.3	2.3	167	Difficult, can be resolved in 10cm telescopes
Gamma (γ) Leporis	05	44.5	−22	27	3.59 / 6.28	95.0	350	White and yellow-orange components, easy pair
Sirius (α Canis Majoris)	06	45.1	−16	43	−1.4 / 8.5			Binary, period 50 years, difficult
Castor (α Geminorum)	07	34.5	+31	53	1.93 / 2.97	7.0	55	Binary, 445 years, widening
Gamma (γ) Velorum	08	09.5	−47	20	1.83 / 4.27	41.2	220	Pretty pair in nice field of stars
Upsilon (υ) Carinae	09	47.1	−65	04	3.08 / 6.10	5.03	129	Nice object in small telescopes
Algieba (γ Leonis)	10	20.0	+19	50	2.28 / 3.51	4.6	126	Binary, 510 years, orange-red and yellow
Acrux (α Crucis)	12	26.4	−63	06	1.40 / 1.90	4.0	114	Glorious pair, third star visible in low power
Porrima (γ Virginis)	12	41.5	−01	27	3.56 / 3.65			Binary, 170 years, widening, visible in small telescopes
Cor Caroli (α Canum Venaticorum)	12	56.0	+38	19	2.90 / 5.60	19.6	229	Easy, yellow and bluish
Mizar (ζ Ursae Majoris)	13	24.0	+54	56	2.3 / 4.0	14.4	152	Easy, wide naked-eye pair with Alcor
Alpha (α) Centauri	14	39.6	−60	50	0.0 / 1.2			Binary, beautiful pair of stars
Izar (ε Boötis)	14	45.0	+27	04	2.4 / 5.1	2.9	344	Fine pair of yellow and blue stars
Omega[1,2] (ω[1,2]) Scorpii	16	06.0	−20	41	4.0 / 4.3	14.6	145	Optical pair, easy
Epsilon[1] (ε[1]) Lyrae	18	44.3	+39	40	4.7 / 6.2	2.6	346	The Double-Double, quadruple system with ε[2]
Epsilon[2] (ε[2]) Lyrae	18	44.3	+39	40	5.1 / 5.5	2.3	76	Both individual pairs just visible in 80mm telescopes
Theta[1,2] (θ[1,2]) Serpentis	18	56.2	+04	12	4.6 / 5.0	22.4	104	Easy pair, mag 6.7 yellow star 7 arc minutes from θ2

Star	RA		Declination		Magnitudes	Separation	PA	Comments
	h	m	°	ʹ		(arcsec)	°	
Albireo (β Cygni)	19	30.7	+27	58	3.1 / 5.1	34.3	54	Glorious pair, yellow and blue-green
Algedi (α¹·² Capricorni)	20	18.0	−12	32	3.7 / 4.3	6.3	292	Optical pair, easy
Gamma (γ) Delphini	20	46.7	+16	07	5.14 / 4.27	9.2	265	Easy, orange and yellow-white
61 Cygni	21	06.9	+38	45	5.20 / 6.05	31.6	152	Binary, 678 years, both orange
Theta (θ) Indi	21	19.9	−53	27	4.6 / 7.2	7.0	275	Fine object for small telescopes
Delta (δ) Tucanae	22	27.3	−64	58	4.49 / 8.7	7.0	281	Beautiful double, white and reddish

Some Interesting Nebulae, Star Clusters and Galaxies

Brian Jones

Object	RA h	RA m	Declination °	Declination '	Remarks
47 Tucanae (in Tucana)	00	24.1	−72	05	Fine globular cluster, easy with naked eye
M31 (in Andromeda)	00	40.7	+41	05	Andromeda Galaxy, visible to unaided eye
Small Magellanic Cloud	00	52.6	−72	49	Satellite galaxy of the Milky Way
NGC 362 (in Tucana)	01	03.3	−70	51	Globular cluster, impressive sight in telescopes
M33 (in Triangulum)	01	31.8	+30	28	Triangulum Spiral Galaxy, quite faint
NGC 869 and NGC 884	02	20.0	+57	08	Sword Handle Double Cluster in Perseus
M34 (in Perseus)	02	42.1	+42	46	Open star cluster near Algol
M45 (in Taurus)	03	47.4	+24	07	Pleiades or Seven Sisters cluster, a fine object
Large Magellanic Cloud	05	23.5	−69	45	Satellite galaxy of the Milky Way
30 Doradus (in Dorado)	05	38.6	−69	06	Star-forming region in Large Magellanic Cloud
M1 (in Taurus)	05	34.5	+22	01	Crab Nebula, near Zeta (ζ) Tauri
M38 (in Auriga)	05	28.6	+35	51	Open star cluster
M42 (in Orion)	05	33.4	−05	24	Orion Nebula
M36 (in Auriga)	05	36.2	+34	08	Open star cluster
M37 (in Auriga)	05	52.3	+32	33	Open star cluster
M35 (in Gemini)	06	06.5	+24	21	Open star cluster near Eta (η) Geminorum
M41 (in Canis Major)	06	46.0	−20	46	Open star cluster to south of Sirius
M44 (in Cancer)	08	38.0	+20	07	Praesepe, visible to naked eye
M81 (in Ursa Major)	09	55.5	+69	04	Bode's Galaxy
M82 (in Ursa Major)	09	55.9	+69	41	Cigar Galaxy or Starburst Galaxy
Carina Nebula (in Carina)	10	45.2	−59	52	NGC 3372, large area of bright and dark nebulosity
M104 (in Virgo)	12	40.0	−11	37	Sombrero Hat Galaxy to south of Porrima
Coal Sack (in Crux)	12	50.0	−62	30	Prominent dark nebula, visible to naked eye
NGC 4755 (in Crux)	12	53.6	−60	22	Jewel Box open cluster, magnificent object
Omega (ω) Centauri	13	23.7	−47	03	Splendid globular in Centaurus, easy with naked eye
M51 (in Canes Venatici)	13	29.9	+47	12	Whirlpool Galaxy
M3 (in Canes Venatici)	13	40.6	+28	34	Bright Globular Cluster

Object	RA		Declination		Remarks
	h	m	°	‘	
M4 (in Scorpius)	16	21.5	−26	26	Globular cluster, close to Antares
M12 (in Ophiuchus)	16	47.2	−01	57	Globular cluster
M10 (in Ophiuchus)	16	57.1	−04	06	Globular cluster
M13 (in Hercules)	16	40.0	+36	31	Great Globular Cluster, just visible to naked eye
M92 (in Hercules)	17	16.1	+43	11	Globular cluster
M6 (in Scorpius)	17	36.8	−32	11	Open cluster
M7 (in Scorpius)	17	50.6	−34	48	Bright open cluster
M20 (in Sagittarius)	18	02.3	−23	02	Trifid Nebula
M8 (in Sagittarius)	18	03.6	−24	23	Lagoon Nebula, just visible to naked eye
M16 (in Serpens)	18	18.8	−13	49	Eagle Nebula and star cluster
M17 (in Sagittarius)	18	20.2	−16	11	Omega Nebula
M11 (in Scutum)	18	49.0	−06	19	Wild Duck open star cluster
M57 (in Lyra)	18	52.6	+32	59	Ring Nebula, brightest planetary
M27 (in Vulpecula)	19	58.1	+22	37	Dumbbell Nebula
M29 (in Cygnus)	20	23.9	+38	31	Open cluster
M15 (in Pegasus)	21	30.1	+12	10	Bright globular cluster near Epsilon (ε) Pegasi
M39 (in Cygnus)	21	31.6	+48	25	Open cluster, good with low powers
M52 (in Cassiopeia)	23	24.2	+61	35	Open star cluster near 4 Cassiopeiae

M = Messier Catalogue Number   NGC = New General Catalogue Number

The positions in the sky of each of the objects contained in this list are given on the Monthly Star Charts printed elsewhere in this volume.

Astronomical Organizations

American Association of Variable Star Observers

49 Bay State Road, Cambridge, Massachusetts 02138, USA

www.aavso.org

The AAVSO is an international non-profit organization of variable star observers whose mission is to enable anyone, anywhere, to participate in scientific discovery through variable star astronomy. We accomplish our mission by carrying out the following activities:

- observation and analysis of variable stars
- collecting and archiving observations for worldwide access
- forging strong collaborations between amateur and professional astronomers
- promoting scientific research, education and public outreach using variable star data

American Astronomical Society

1667 K Street NW, Suite 800, Washington, DC 20006, USA

https://aas.org

Established in 1899, the American Astronomical Society (AAS) is the major organization of professional astronomers in North America. The mission of the AAS is to enhance and share humanity's scientific understanding of the universe, which it achieves through publishing, meeting organization, education and outreach, and training and professional development.

Astronomical Society of the Pacific

390 Ashton Avenue, San Francisco, CA 94112, USA

www.astrosociety.org

Formed in 1889, the Astronomical Society of the Pacific (ASP) is a non-profit membership organization which is international in scope. The mission of the ASP is to increase the understanding and appreciation of astronomy through the engagement of our many constituencies to advance science and science literacy. We invite you to explore our site to learn more about us; to check out our resources and education section for the researcher, the educator, and the backyard enthusiast; to get involved by becoming an ASP member; and to consider supporting our work for the benefit of a science literate world!

Astrospeakers.org
www.astrospeakers.org
A website designed to help astronomical societies and clubs locate astronomy and space lecturers which is also designed to help people find their local astronomical society. It is completely free to register and use and, with over 50 speakers listed, is an excellent place to find lecturers for your astronomical society meetings and events. Speakers and astronomical societies are encouraged to use the online registration to be added to the lists.

British Astronomical Association
Burlington House, Piccadilly, London, W1J 0DU, England
www.britastro.org
The British Astronomical Association is the UK's leading society for amateur astronomers catering for beginners to the most advanced observers who produce scientifically useful observations. Our Observing Sections provide encouragement and advice about observing. We hold meetings around the country and publish a bi-monthly Journal plus an annual Handbook. For more details, including how to join the BAA or to contact us, please visit our website.

British Interplanetary Society
Arthur C Clarke House, 27/29 South Lambeth Road, London, SW8 1SZ, England
www.bis-space.com
The British Interplanetary Society is the world's longest-established space advocacy organisation, founded in 1933 by the pioneers of British astronautics. It is the first organisation in the world still in existence to design spaceships. Early members included Sir Arthur C Clarke and Sir Patrick Moore. The Society has created many original concepts, from a 1938 lunar lander and space suit designs, to geostationary orbits, space stations and the first engineering study

of a starship, Project Daedalus. Today the BIS has a worldwide membership and welcomes all with an interest in Space, including enthusiasts, students, academics and professionals.

Canadian Astronomical Society
Société Canadienne D'astronomie (CASCA)
100 Viaduct Avenue West, Victoria, British Columbia, V9E 1J3, Canada
www.casca.ca
CASCA is the national organization of professional astronomers in Canada. It seeks to promote and advance knowledge of the universe through research and education. Founded in 1979, members include university professors, observatory scientists, postdoctoral fellows, graduate students, instrumentalists, and public outreach specialists.

Royal Astronomical Society of Canada
203-4920 Dundas St W, Etobicoke, Toronto, ON M9A 1B7, Canada
www.rasc.ca
Bringing together over 5,000 enthusiastic amateurs, educators and professionals RASC is a national, non-profit, charitable organization devoted to the advancement of astronomy and related sciences and is Canada's leading astronomy organization. Membership is open to everyone with an interest in astronomy. You may join through any one of our 29 RASC centres, located across Canada and all of which offer local programs. The majority of our events are free and open to the public.

Federation of Astronomical Societies
The Secretary, 147 Queen Street, SWINTON, Mexborough, S64 8NG
www.fedastro.org.uk
The Federation of Astronomical Societies (FAS) is an umbrella group for astronomical societies in the UK. It promotes cooperation, knowledge and information sharing and encourages best practice. The FAS aims to be a body of societies united in their attempts to help each other find the best ways of working for their common cause of creating a fully successful astronomical society. In this way it endeavours to be a true federation, rather than some remote central organization disseminating information only from its own

limited experience. The FAS also provides a competitive Public Liability Insurance scheme for its members.

International Dark-Sky Association
darksky.org

The International Dark-Sky Association (IDA) is the recognized authority on light pollution and the leading organization combating light pollution worldwide. The IDA works to protect the night skies for present and future generations, our public outreach efforts providing solutions, quality education and programs that inform audiences across the United States of America and throughout the world. At the local level, our mission is furthered through the work of our U.S. and international chapters representing five continents.

The goals of the IDA are:

- Advocate for the protection of the night sky
- Educate the public and policymakers about night sky conservation
- Promote environmentally responsible outdoor lighting
- Empower the public with the tools and resources to help bring back the night

Royal Astronomical Society of New Zealand
PO Box 3181, Wellington, New Zealand
www.rasnz.org.nz

Founded in 1920, the object of The Royal Astronomical Society of New Zealand is the promotion and extension of knowledge of astronomy and related branches of science. It encourages interest in astronomy and is an association of observers and others for mutual help and advancement of science. Membership is open to all interested in astronomy. The RASNZ has about 180 individual members including both professional and amateur astronomers and many of the astronomical research and observing programmes carried out in New Zealand involve collaboration between the two. In addition the society has a number of groups or sections which cater for people who have interests in particular areas of astronomy.

Astronomical Society of Southern Africa

Astronomical Society of Southern Africa, c/o SAAO, PO Box 9, Observatory, 7935, South Africa

assa.saao.ac.za

Formed in 1922, The Astronomical Society of Southern Africa comprises both amateur and professional astronomers. Membership is open to all interested persons. Regional Centres host regular meetings and conduct public outreach events, whilst national Sections coordinate special interest groups and observing programmes. The Society administers two Scholarships, and hosts occasional Symposia where papers are presented. For more details, or to contact us, please visit our website.

Royal Astronomical Society

Burlington House, Piccadilly, London, W1J 0BQ, England

www.ras.org.uk

The Royal Astronomical Society, with around 4,000 members, is the leading UK body representing astronomy, space science and geophysics, with a membership including professional researchers, advanced amateur astronomers, historians of science, teachers, science writers, public engagement specialists and others.

Society for Popular Astronomy

Secretary: Guy Fennimore, 36 Fairway, Keyworth, Nottingham, NG12 5DU

www.popastro.com

The Society for Popular Astronomy is a national society that aims to present astronomy in a less technical manner. The bi-monthly society magazine Popular Astronomy is issued free to all members.

Our Contributors

Martin Beech is Professor of Astronomy at Campion College, one of the Federated Colleges at the University of Regina in Saskatchewan, Canada. His main astronomical research interests have focused on small solar system bodies: asteroids and comets, along with meteoroid streams, impact cratering, meteorites and atmospheric ablation effects. He has published several books relating to terrestrial catastrophism, terraforming and the search for extraterrestrial mega-structures. He has additionally written extensively on the history of science and specifically the history of scientific instruments. As a past managing editor and contributor to the UK's *Astronomy Now* magazine, he has long held an active interest in the promotion of astronomy to the general public.

Steve Brown is an amateur astronomer based in Stokesley in the North Riding of Yorkshire. He has been interested in astronomy from a young age and has observed seriously since buying his first telescope (a 130mm reflector) in 2011. He regularly observes from his garden (when the Yorkshire weather allows) dividing his observing time between astrophotography and sketching. His image 'The Rainbow Star' won the Stars and Nebulae category of the Insight Astronomy Photographer of the Year competition in 2016. Steve also appeared on the Sky at Night episode 'Moore Winter Marathon', filmed at Kielder Observatory and broadcast in March 2013. The programme also featured several of his astro sketches. Steve has also had sketches, images and articles published in *Sky at Night* and *All About Space* magazines, as well as in a number of astronomy books. As a member of the British Astronomical Association, Steve regularly submits meteor shower observations to their Meteor Section. You can follow Steve on Twitter via @**sjb_astro** and see his images on Flickr at **www.flickr.com/photos/sjb_astro**

Mike Frost is a Fellow of the Royal Astronomical Society and Director of the British Astronomical Association's historical section. His research interests centre on astronomers with connections to the English Midlands where he lives, and he lectures about them to astronomy societies around the country. Mike's day job is as a systems engineer, designing and commissioning control systems for steel rolling mills around the world.

Neil Haggath has a degree in astrophysics from Leeds University and has been a Fellow of the Royal Astronomical Society since 1993. A member of Cleveland and Darlington Astronomical Society for 37 years, he has served on its committee for 29 years. Neil is an avid umbraphile, clocking up six total eclipse expeditions so far, to locations as far flung as Australia and Hawaii. Four of them were successful, the most recent being in Jackson, Wyoming on 21 August 2017. In 2012, he may have set a somewhat unenviable record among British astronomers – for the greatest distance travelled (6,000 miles to Thailand) to NOT see the transit of Venus. He saw nothing on the day … and got very wet!

Dr. David M. Harland gained his BSc in astronomy in 1977, lectured in computer science, worked in industry, and managed academic research. In 1995 he 'retired' in order to write on space themes.

David Harper, FRAS has had a varied career which includes teaching mathematics, astronomy and computing at Queen Mary University of London, astronomical software development at the Royal Greenwich Observatory, bioinformatics support at the Wellcome Trust Sanger Institute, and a research interest in the dynamics of planetary satellites, which began during his Ph.D. at Liverpool University in the 1980s and continues in an occasional collaboration with colleagues in China. He is married to fellow contributor Lynne Stockman.

Tracie Heywood is an amateur astronomer from Leek in Staffordshire and is one of the UK's leading variable star observers, using binoculars to monitor the brightness changes of several hundred variable stars. Tracie currently writes a monthly column about variable stars for *Astronomy Now* magazine. She has previously been the Eclipsing Binary coordinator for the Variable Star Section of the British Astronomical Association and the Director of the Variable Star Section of the Society for Popular Astronomy.

Rod Hine was aged around ten when he was given a copy of *The Boys Book of Space* by Patrick Moore. Already interested in anything to do with science and engineering he devoured the book from cover to cover. The launch of Sputnik I shortly afterwards clinched his interest in physics and space travel. He took physics, chemistry and mathematics at A-level and then studied Natural Sciences at Churchill College, Cambridge. He later switched to Electrical Sciences and subsequently joined Marconi at Chelmsford working on satellite communications. This led to work in meteorological communications in Nairobi, Kenya and later a teaching post at the Kenya Polytechnic. There he met and married a Yorkshire lass and moved back to the UK in 1976. Since then he has had a variety of jobs in electronics and industrial controls, and recently has been lecturing part-time at the University of Bradford. Rod got fully back into astronomy in around 1992 when his wife bought him an astronomy book, at which time he joined Bradford Astronomical Society.

Brian Jones hails from Bradford in the West Riding of Yorkshire and was a founder member of the Bradford Astronomical Society. He developed a fascination with astronomy at the age of five when he first saw the stars through a pair of binoculars, although he spent the first part of his working life developing a career in mechanical engineering. However, his true passion lay in the stars and his interest in astronomy took him into the realms of writing sky guides for local newspapers, appearing on local radio and television, teaching astronomy and space in schools and, in 1985, leaving engineering to become a full time astronomy and space writer. His books have covered a range of astronomy and space-related topics for both children and adults and his journalistic work includes writing articles and book reviews for several astronomy magazines as well as for many general interest magazines, newspapers and periodicals. His passion for bringing an appreciation of the universe to his readers is reflected in his writing.

You can follow Brian on Twitter via **@StarsBrian** and check out the sky by visiting his blog at **starlight-nights.co.uk** from where you can also access his Facebook group Starlight Nights.

Carolyn Kennett lives in the far south west of Cornwall. She likes to write, although you will often find her walking the Tinner's Way and coastal pathways found in the local countryside. As well as researching local astronomy history, she has a passion for archaeoastronomy and 18th and 19th century maritime astronomy. She currently co-edits *Bulletin*, the magazine of the Society for the History of Astronomy. Further details of the SHA can be found at **societyforthehistoryofastronomy.com**

Bill Leatherbarrow is Emeritus Professor of Russian Studies at the University of Sheffield and a lifelong amateur astronomer. He is an active observer of the Moon and planets and has contributed to previous editions of the *Yearbook of Astronomy*. He served as President of the British Astronomical Association (BAA) from 2011 to 2013 and is currently Director of the BAA Lunar Section. He writes regularly for the magazine *Astronomy Now* and is the author of the book *The Moon* (Reaktion Books, 2018).

John McCue graduated in astronomy from the University of St Andrews and began teaching. He gained a Ph.D. from Teesside University studying the unusual rotation of Venus. In 1979 he and his colleague John Nichol founded the Cleveland and Darlington Astronomical Society, which then worked in partnership with the local authority to build the Wynyard Planetarium and Observatory in Stockton-on-Tees. John is currently double star advisor for the British Astronomical Association.

Bob Mizon graduated from King's College London in 1969 with a degree in Modern Languages. His abiding interest in astronomy was sparked by a chance encounter at the age of ten with the *Larousse Encyclopaedia of Astronomy* in a primary school library. This culminated in his switching in 1996 from teaching French to operating a mobile schools' planetarium and speaking to groups and societies all over the UK. Bob is a Fellow of the Royal Astronomical Society and was awarded an MBE in 2010 for his work as an astronomy educator and as coordinator of the British Astronomical Association's Commission for Dark Skies (CfDS) which can be seen at **www.britastro.org/dark-skies**. A Lifetime Member of the International Dark-Sky Association (IDA), Bob has worked with CfDS colleagues in support of several of Britain's IDA Dark Sky Reserves and Parks.

Neil Norman, FRAS first became fascinated with the night sky when he was five years of age and saw Patrick Moore on the television for the first time. It was the Sky at Night programme, broadcast in March 1986 and dedicated to the

Giotto probe reaching Halley's Comet, which was to ignite his passion for these icy interlopers. As the years passed, he began writing astronomy articles for local news magazines before moving into internet radio where he initially guested on the Astronomyfm show 'Under British Skies', before becoming a co-host for a short time. In 2013 he created Comet Watch, a Facebook group dedicated to comets of the past, present and future. His involvement with Astronomyfm led to the creation of the monthly radio show 'Cometwatch', which is now in its second year. Neil lives in Suffolk with his partner and three children.

Julian Onions has had an interest in astronomy over many years, but decided to take it further by studying for a doctorate in astrophysics at Nottingham University. There he studies computer models of galaxy formation using some of the biggest computers in the world, and now builds model universes using mostly dark matter. Julian is also a keen amateur astrophotographer, taking images of various astronomical objects. However, he is brought back to earth by his department colleagues, who have access to some of the largest telescopes yet built.

Roger Pickard has been observing variable stars for more years than he cares to remember. Initially this was as a visual observer, but he then dabbled in photoelectric photometry (PEP) just before the advent of affordable CCDs. He now operates his telescope and CCD camera, which is located at the end of his garden, from within the warmth of his house. Roger has been the Director of the British Astronomical Association Variable Star Section since 1999 and can be contacted at **roger.pickard@sky.com** by anyone who needs help with variable star observing.

Sian Prosser has been Librarian and Archivist at the Royal Astronomical Society since 2014, managing a remarkable astronomy and geophysics collection and helping people to access and make use of it. A language degree at the University of Glasgow and an MA in Medieval Studies at the University of Leeds was followed by roles in logistics and export. Sian then joined the French department of the University of Sheffield for an AHRC-funded PhD on what medieval manuscripts can tell us about how the Troy legend was understood in 13th-century France. Her first library post was at the Brotherton Library, University of Leeds from where she went on to manage a small team at University of

Warwick Library. Sian completed the MA in Library and Information Studies at University College London in 2011, and has returned there to study for the Certificate of Higher Education in Astronomy.

Greg Quicke has been delivering intensely practical astronomy to humanity since 1995 through live performance under a real sky using big telescopes. With over a hundred thousand people joining him over that time, the BBC eventually found him and enlisted him as their practical astronomer to work with Professor Brian Cox on Stargazing Live. Greg has learned his stars through a life lived outdoors in the wild Kimberley region of Western Australia. Working as a Pearl Shell diver, he learned first hand how the sun, the moon and the earth interact to create the 10m tides of his hometown of Broome. This was his first astronomical step on a lifelong journey of discovery that continues to inspire, educate and entertain both himself and those who join him under the stars. Greg performs at the Astro Tours Dark Sky Launchpad outside of Broome on set dates from April through October when the skies are clear. He also speaks and performs at corporate and festival events across Australia and at occasional international events. Links to his TV series, live performance dates and all bookings can be found at **www.astrotours.net**

Peter Rea has had a keen interest in lunar and planetary exploration since the early 1960s and frequently lectures on the subject. He helped found the Cleethorpes and District Astronomical Society in 1969. In April of 1972 he was at the Kennedy Space Centre in Florida to see the launch of Apollo 16 to the moon and in October 1997 was at the southern end of Cape Canaveral to see the launch of Cassini to Saturn. He would still like to see a total solar eclipse as the expedition he was on to see the 1973 eclipse in Mali had vehicle trouble and the meteorologists decided he was not going to see the 1999 eclipse from Devon. He lives in Lincolnshire with his wife Anne and has a daughter who resides in Melbourne, Australia.

Courtney Seligman is an Emeritus Professor of Astronomy at Long Beach City College, California where he taught astronomy for more than forty years and presented telescope and planetarium shows as a public service. He maintains and is still working on an astronomy website at **cseligman.com** which contains an online astronomy text, a sky atlas and a catalogue of the NGC/IC and other astronomical objects. More information about him and his interests is available at **cseligman.com/about.htm**

Lynne Marie Stockman holds degrees in mathematics from Whitman College, the University of Washington and the University of London, and has studied astronomy at both undergraduate and postgraduate levels. She is a native of North Idaho but has lived in Britain for the past 28 years. Lynne was an early pioneer of the World Wide Web: with her husband David Harper, she created the web site **obliquity.com** in 1998 to share their interest in astronomy, computing, family history and cats.

Susan Stubbs first became interested in stars as a child when she was astonished by the appearance of the Milky Way after seeing it from a dark sky site in Sussex whilst camping with the Guides. However, progressing to her Guide Stargazers badge showed her that not so many people were interested in astronomy back then, as finding an enthusiast to test her proved problematic! University, career and family then got in the way for quite a few years, but she returned to the fold via Open University astronomy courses and through joining Bradford Astronomical Society of which she is an active member. She was delighted to find that so many more people are interested in astronomy now than all those years ago in childhood. Susan recently had a new astronomical experience, travelling with her husband Robin to see the August 2017 total solar eclipse in Wyoming. She definitely wants to do it again!

Martin Whipp has been interested in astronomy since the age of 5 and joined the York Astronomical Society when he was just 14. Over the years he has held most positions on their committee, including that of Chairman a number of times. Martin was made a fellow of the Royal Astronomical Society in 2001. He specialises in time lapse photography and his film work has been used in several television productions, including *The Sky at Night*. He also enjoys public outreach, including giving lectures at numerous public events hosted by York Astronomical Society. He recently took on the role of presenter at the newly built Lime Tree Observatory, situated near Grewelthorpe in the Yorkshire Dales, which is home to a 24-inch reflecting telescope built by none other than John Wall, inventor of the Crayford focuser.

Glossary

Brian Jones and David Harper

Airburst

The violent 'explosion' and resulting energy shockwave of a small **asteroid** or **meteorite** which has entered the Earth's atmosphere, and which occurs before the object reaches the ground. The Tunguska event of 1908 is believed to have been an asteroid airburst.

Altitude

The altitude of a star or other object is its angular distance above the horizon. For example, if a star is located at the **zenith**, or overhead point, its altitude is 90° and if it is on the horizon, its altitude is 0°.

Angular Distance

The angular distance between two objects on the sky is the angle subtended between the directions to the two objects, either at the centre of the Earth (geocentric angular distance) or the observer's eye (apparent angular distance). It is most commonly expressed in degrees, or for smaller angular distances, minutes of arc or seconds of arc.

Antoniadi Scale

A scale of seeing conditions named after astronomer Eugène Michel Antoniadi who devised it during the early 1900s. It assesses the weather and seeing conditions under which astronomical observations are carried out. The Antoniadi scale has five gradations, these being: (I) perfect seeing with no quivering; (II) good seeing, some slight undulations with frequent steady moments; (III) moderate seeing, about equal steady and turbulent moments; (IV) poor seeing, with constant undulations making sketching difficult; and (V) very bad seeing, with turbulence scarcely allowing a sketch to be made.

Aphelion

This is the point at which an object, such as a planet, comet or asteroid travelling in an elliptical **orbit**, is at its maximum distance from the Sun.

Apogee

This is the point in its **orbit** around the Earth at which an object is at its furthest from the Earth.

Apparition

The period during which a planet is visible, usually starting at **conjunction** with the Sun, running through **opposition** (for a superior planet) or **greatest elongation** (for Mercury or Venus), and ending with the next conjunction with the Sun.

Appulse
The close approach, as seen from the Earth, between two planets, or a planet and a star, or the Moon and a star or planet. Also known as a *conjunction*.

Asterism
An asterism is grouping or collection of stars often (but not always) located within a *constellation* that forms an apparent and distinctive pattern in its own right. Well known examples include the Plough (in Ursa Major); the False Cross (formed from stars in Carina and Vela); and the Summer Triangle, which is formed from the bright stars Vega (in Lyra), Deneb (in Cygnus) and Altair (in Aquila).

Asteroid
Another name for a *minor planet*.

Autumnal Equinox
The autumnal equinox is the point at which the apparent path of the Sun, moving from north to south, crosses the *celestial equator*. In the Earth's northern hemisphere this marks the start of autumn, whilst in the southern hemisphere it is the start of spring.

Averted Vision
Averted vision is a useful technique for observing faint objects which involves looking slightly to one side of the object under observation and, by doing so, allowing the light emitted by the object to fall on the part of the retina that is more sensitive at low light levels. Although you are not looking directly at the object, it is surprising how much more detail comes into view. This technique is also useful when observing double stars which have components of greatly contrasting brightness. Although direct vision may not reveal the glow of a faint companion star in the glare of a much brighter primary, averted vision may well bring the fainter star into view.

Azimuth
The azimuth of a star or other object is its angular position measured round the *horizon* from north (azimuth 0°) through east (azimuth 90°), south (azimuth 180°) and west (azimuth 270°). The azimuth and *altitude*, taken together, define the position of the object referred to the observer's *local horizon*.

Barycentre
The barycentre is the centre of mass of two or more bodies that are orbiting each other (such as a planet and satellite or two components of a *binary star* system) and which is therefore the point around which they both *orbit*.

Binary Star
See Double Star

Black Hole
A region of space around a very compact and extremely massive collapsed star within which the gravitational field is so intense that not even light can escape.

Caldwell Catalogue
This is a catalogue of 109 star clusters, nebulae, and galaxies compiled by Patrick Moore to complement the *Messier Catalogue*. Intended for use as an observing guide by amateur astronomers it includes a number of bright *deep sky objects* that did not find their way into the Messier Catalogue, which was originally compiled as a list of known objects that might be confused with comets. Moore used his other surname (Caldwell) to name the list and the objects within it (the first letter of 'Moore' having been used for the Messier Catalogue) and entries in the Caldwell Catalogue are designated with a 'C' followed by the catalogue number (1 to 109).

Amongst the 109 objects in the Caldwell Catalogue are the Sword Handle Double Cluster NGC 869 and NGC 884 (C14) in Perseus; supernova remnant(s) the East Veil Nebula and West Veil Nebula (C33 and C34) in Cygnus; the Hyades open star cluster (C41) in Taurus; and Hubble's Variable Nebula (C46) in Monoceros. Unlike the Messier Catalogue, which was compiled from observations made by Charles Messier from Paris, the Caldwell Catalogue contains deep sky objects visible from the southern hemisphere, such as the Centaurus A galaxy (C77) and globular cluster Omega Centauri (C80) in Centaurus; the Jewel Box open star cluster (C94) in Crux and the globular cluster 47 Tucanae (C106) in Tucana.

Although few of the objects detailed elsewhere in the Yearbook of Astronomy carry a Caldwell Catalogue reference, it was felt that an entry should appear in the Glossary as the catalogue is nonetheless an important guide for the backyard astronomer.

Celestial Equator
The celestial equator is a projection of the Earth's *equator* onto the *celestial sphere*, equidistant from the *celestial poles* and dividing the celestial sphere into two hemispheres.

Celestial Poles
The north (and south) celestial poles are points on the *celestial sphere* directly above the north and south terrestrial poles around which the celestial sphere appears to rotate and through which extensions of the Earth's axis of rotation would pass.

The north celestial pole, the position of which is at marked at present by the relatively bright star Polaris, lies in the constellation Ursa Minor (the Little Bear) and would be seen directly overhead when viewed from the North Pole. There is no particularly bright star marking the position of the south celestial pole, which lies in the tiny *constellation* Octans (the Octant) and which would be situated directly overhead when seen from the South Pole. The north celestial pole lies in the direction of north when viewed from elsewhere on the Earth's surface and the south celestial pole lies in the direction of south when viewed from other locations.

Celestial Sphere
The imaginary sphere surrounding the Earth on which the stars appear to lie.

Circumpolar Star

A circumpolar star is a star which never sets from a given **latitude**. When viewing the sky from either the North or South Pole, all stars will be circumpolar, although no stars are circumpolar when viewed from the equator.

Comet

A comet is an object comprised of a mixture of gas, dust and ice which travels around the Sun in an orbit that can often be very eccentric.

Conjunction

This is the position at which two objects are lined up with each other (or nearly so) as seen from Earth. Superior conjunction occurs when a planet is at the opposite side of the Sun as seen from Earth and inferior conjunction when a planet lies between the Sun and Earth.

Constellation

A constellation is an arbitrary grouping of stars forming a pattern or imaginary picture on the celestial sphere. Many of these have traditional names and date back to ancient Greece or even earlier and are associated with the folklore and mythology of the time. There are also some of what may be described as 'modern' constellations, devised comparatively recently by astronomers during the last few centuries. There are 88 official constellations which together cover the entire sky, each one of which refers to and delineates that particular region of the **celestial sphere**, the result being that every celestial object is described as being within one particular constellation or another.

Dark Nebula

See Nebula.

Declination

This is the **angular distance** between a celestial object and the celestial equator. Declination is expressed in degrees, minutes and seconds either north (N) or south (S) of the **celestial equator**.

Deep Sky Object

Deep sky objects are objects (other than individual stars) which lie beyond the confines of our **Solar System**. They may be either galactic or extra-galactic and include such things as **star clusters**, **nebulae** and **galaxies**.

Direct Motion

A planet is in direct (or prograde) motion when its **right ascension** or ecliptic **longitude** is increasing with the passing of time. This means that it is moving eastwards with respect to the background stars.

Double Stars

Double stars are two stars which appear to be close together in space. Although some double stars (known as *optical* doubles) are made up of two stars that only happen to lie in the same line of sight as seen from Earth and are nothing more than chance alignments, most are comprised of stars that are gravitationally linked and orbit each other, forming a genuine double-star system (also known as a *binary* star).

Eclipse

An eclipse is the obscuration of one celestial object by another, such as the Sun by the Moon during a solar eclipse or one component of an eclipsing **binary star** by the companion star.

* A **solar eclipse** occurs when the Moon passes directly between the Earth and the Sun. There are three types of solar eclipse. A total solar eclipse takes place when the Moon completely obscures the Sun, during which event the Sun's corona, or outer atmosphere, is revealed; a partial solar eclipse occurs when the lining up of the Earth, Moon and Sun is not exact and the Moon covers only a part of the Sun; an annular solar eclipse takes place when the Moon is at or near its farthest from Earth, at which time the lunar disc appears smaller and does not completely cover the solar disc, the Sun's visible outer edges forming a 'ring of light' or 'annulus' around the Moon. Some eclipses which begin as annular may become total along part of their path; these are known as hybrid eclipses, and are quite rare.
* A **lunar eclipse** occurs when the Earth passes between the Sun and the Moon, and the Earth's shadow is thrown onto the lunar surface. There are three types of lunar eclipse. A total lunar eclipse takes place when the Moon passes completely through the **umbra** of the Earth's shadow, during which process the Moon will gradually darken and take on a reddish/rusty hue; a partial lunar eclipse occurs when the Moon passes through the **penumbra** of the Earth's shadow and only part of it enters the umbra; a penumbral lunar eclipse takes place when the Moon only enters the penumbra of the Earth's shadow without touching or entering the umbra.

Ecliptic

As the Earth orbits the Sun, its position against the background stars changes slightly from day to day, the overall effect of this being that the Sun appears to travel completely around the **celestial sphere** over the course of a year. The apparent path of the Sun is known as the ecliptic and is superimposed against the band of **constellations** we call the **Zodiac** through which the Sun appears to move.

Ellipse

The closed, oval-shaped form obtained by cutting through a cone at an angle to the main axis of the cone. The orbits of the planets around the Sun are all elliptical.

Elongation (and Greatest Elongation)

In its most general sense, elongation refers to the angular separation between two celestial objects as seen from a third object. It is most often used to refer to the **angular distance** between the Sun and a planet or the Moon, as seen from the Earth.

The greatest elongation of Mercury or Venus is the maximum angular distance between the planet and the Sun as seen from the Earth, during a particular *apparition*.

Emission Nebula
See Nebula.

Ephemeris (plural: Ephemerides)
Table showing the predicted positions of celestial objects such as comets or planets.

Equator
The equator of a planet or other spheroidal celestial body is the great circle on the surface of the body whose latitude is zero, as defined by the axis of rotation. The *celestial equator* is the projection of the plane of the Earth's equator onto the sky.

Equinox
The equinoxes are the two points at which the ecliptic crosses the *celestial equator* (see also *Autumnal Equinox* and *Vernal Equinox*). The term is also used to denote the dates on which the Sun passes these points on the *ecliptic*.

Exoplanet
An exoplanet (or extrasolar planet) is a planet orbiting a star outside of our *Solar System*.

Galaxy
A galaxy is a vast collection of stars, gas and dust bound together by gravity and measuring many thousands of light years across. Galaxies occur in a wide variety of shapes and sizes including spiral, elliptical and irregular and most are so far away that their light has taken many millions of years to reach us. Our *Solar System* is situated in the Milky Way Galaxy, a spiral galaxy containing several billion stars. Located within the *Local Group of Galaxies*, the *Milky Way* Galaxy is often referred to simply as the Galaxy.

Horizon
The horizon is a great circle that is theoretically defined by a zenith distance of 90 degrees. In practice, the observer's *local horizon* will differ from this.

Index Catalogue (IC)
References such as that for IC 2391 (in Vela) and IC 2602 (in Carina) are derived from their numbers in the Index Catalogue (IC), published in 1895 as the first of two supplements (the second was published in 1908) to his *New General Catalogue* of Nebulae and Clusters of Stars (NGC) by the Danish astronomer John Louis Emil Dreyer (1852–1926). Between them, the two Index Catalogues contained details of an additional 5,386 objects.

Inferior Planet
An inferior planet is a planet that travels around the Sun inside the *orbit* of the Earth.

International Astronomical Union (IAU)

Formed in 1919 and based at the Institut d'Astrophysique de Paris, this is the main coordinating body of world astronomy. Its main function is to promote, through international cooperation, all aspects of the science of astronomy. It is also the only authority responsible for the naming of celestial objects and the features on their surfaces.

Latitude

The latitude of the Sun, Moon or planet is its angular distance above or below the *ecliptic*. Note that the *angular distance* of a celestial body north or south of the *celestial equator* is called *declination*, and not latitude.

The latitude of a point on the Earth's surface is its angular distance north or south of the *equator*.

Light Year

To express distances to the stars and other galaxies in miles would involve numbers so huge that they would be unwieldy. Astronomers therefore use the term 'light year' as a unit of distance. A light year is the distance that a beam of light, travelling at around 300,000 km (186,000 miles) per second, would travel in a year and is equivalent to just under 10 trillion km (6 trillion miles).

Local Group of Galaxies

This is a gravitationally-bound collection of galaxies which contains over 50 individual members, one of which is our own Milky Way Galaxy. Other members include the Large Magellanic Cloud, the Small Magellanic Cloud, the Andromeda Galaxy (M31), the Triangulum Spiral Galaxy (M33) and many others.

Galaxies are usually found in groups or clusters. Apart from our own Local Group, many other groups of galaxies are known, typically containing anywhere up to 50 individual members. Even larger than the groups are clusters of galaxies which can contain hundreds or even thousands of individual galaxies. Groups and clusters of galaxies are found throughout the universe.

Local Horizon

The horizon seen by an observer on land or at sea differs from the ideal theoretical horizon, defined as 90 degrees from the *zenith*, due to several factors. This can affect astronomical observations. On land, distant features such as mountains may delay the appearance of the rising Sun, Moon or stars by minutes or even hours compared to rising times tabulated in almanacs. At sea, altitudes measured relative to the sea horizon are affected by the observer's height above sea level. At a height of 30 metres above sea level (an aircraft carrier deck, for example), this 'dip' of the sea horizon is 10 arc-minutes, and the *altitude* of a star observed using a nautical sextant must have this amount subtracted before it can be used to determine position at sea. The effect may seem small, but 1 arc-minute of observed altitude corresponds to one nautical mile, so ignoring the 10 arc-minute dip correction would lead to an error of 10 nautical miles in the position of the ship.

Local Hour Angle

The local hour angle of a star or other celestial object is the difference between the local *sidereal time* and the object's *right ascension*. At upper *transit*, an object's local hour angle is zero. Before transit, the local hour angle is negative, whilst after transit, it is positive.

Longitude

The longitude of the Sun, Moon or planet is its angular position, measured along the *ecliptic* from the First Point of Aries.

The longitude of a point on the Earth's surface is its *angular distance* east or west of the *prime meridian* through Greenwich. By convention, terrestrial longitude is positive east of Greenwich and negative west of Greenwich.

Lunar

Of or appertaining to the Moon.

Lunar Eclipse

See Eclipse.

Magnitude

The magnitude of a star is purely and simply a measurement of its brightness. In around 150BC the Greek astronomer Hipparchus divided the stars up into six classes of brightness, the most prominent stars being ranked as first class and the faintest as sixth. This system classifies the stars and other celestial objects according to how bright they actually appear to the observer. In 1856 the English astronomer Norman Robert Pogson refined the system devised by Hipparchus by classing a 1st magnitude star as being 100 times as bright as one of 6th magnitude, giving a difference between successive magnitudes of $\sqrt[5]{100}$ or 2.512. In other words, a star of magnitude 1.00 is 2.512 times as bright as one of magnitude 2.00, 6.31 (2.512 x 2.512) times as bright as a star of magnitude 3.00 and so on. The same basic system is used today, although modern telescopes enable us to determine values to within 0.01 of a magnitude or better. Negative values are used for the brightest objects including the Sun (−26.8), Venus (−4.4 at its brightest) and Sirius (−1.46). Generally speaking, the faintest objects that can be seen with the naked eye under good viewing conditions are around 6th magnitude, with binoculars allowing you to see stars and other objects down to around 9th magnitude.

Meridian

This is a great circle crossing the *celestial sphere* and which passes through both *celestial poles* and the *zenith*.

Messier Catalogue and References

References such as that for Messier 1 (M1) in Taurus, Messier 31 (M31) in Andromeda and Messier 57 (M57) in Lyra relate to a range of deep sky objects derived from the *Catalogue des Nébuleuses et des Amas d'Étoiles* (Catalogue of Nebulae and Star Clusters) drawn up by the French astronomer Charles Messier during the latter part of the eighteenth century.

Meteor

This is a streak of light in the sky seen as the result of the destruction through atmospheric friction of a *meteoroid* in the Earth's atmosphere.

Meteorite

A meteorite is a *meteoroid* which is sufficiently large to at least partially survive the fall through Earth's atmosphere.

Meteoroid

This is a term applied to particles of interplanetary meteoritic debris

Milky Way

This is the name given to the faint pearly band of light that we sometimes see crossing the sky and which is formed from the collective glow of the combined light from the thousands of stars that lie along the main plane of our Galaxy as seen from Earth. The vast majority of these stars are too faint to be seen individually without some form of optical aid. However, provided the sky is really dark and clear, the Milky Way itself is easily visible to the unaided eye, and any form of optical aid will show that it is indeed made up of many thousands of individual stars. Our *Solar System* lies within the main plane of the Milky Way Galaxy and is located inside one of its spiral arms. The Milky Way is actually our view of the Galaxy, looking along the main galactic plane. The glow we see is the combined light from many different stars and is visible as a continuous band of light stretching completely around the *celestial sphere*.

Nadir

This is the point on the *celestial sphere* directly opposite the *zenith*.

Nebula

Nebulae are huge interstellar clouds of gas and dust. Observed in other galaxies as well as our own, their collective name is from the Latin '*nebula*' meaning 'mist' or 'vapour', and there are three basic types:

- **Emission nebulae** contain young, hot stars that emit copious amounts of ultra-violet radiation which reacts with the gas in the nebula causing the nebula to shine at visible wavelengths and with a reddish colour characteristic of this type of nebula. In other words, emission nebulae *emit* their own light. A famous example is the Orion Nebula (M42) in the constellation Orion which is visible as a shimmering patch of light a little to the south of the three stars forming the Belt of Orion.
- The stars that exist in and around **reflection nebulae** are not hot enough to actually cause the nebula to give off its own light. Instead, the dust particles within them simply *reflect* the light from these stars. The stars in the Pleiades star cluster (M45) in Taurus are surrounded by reflection nebulosity. Photographs of the Pleiades cluster show the nebulosity as a blue haze, this being the characteristic colour of reflection nebulae.

- **Dark nebulae** are clouds of interstellar matter which contain no stars and whose dust particles simply blot out the light from objects beyond. They neither emit or reflect light and appear as dark patches against the brighter backdrop of stars or nebulosity, taking on the appearance of regions devoid of stars. A good example is the Coal Sack in the constellation Crux, a huge blot of matter obscuring the star clouds of the southern Milky Way.

Neutron Star
This is the remnant of a massive star which has exploded as a *supernova*.

New General Catalogue (NGC)
References such as that for NGC 869 and NGC 884 (in Perseus) and NGC 4755 (in Crux) are derived from their numbers in the New General Catalogue of Nebulae and Clusters of Stars (NGC) first published in 1888 by the Danish astronomer John Louis Emil Dreyer (1852–1929) and which contains details of 7,840 star clusters, nebulae and galaxies.

Occultation
This is the temporary covering up of one celestial object, such as a star, by another, such as the Moon or a planet.

Opposition
Opposition is the point in the orbit of a *superior planet* when it is located directly opposite the Sun in the sky.

Orbit
This is the path of one object around another under the influence of gravity.

Parallax
Parallax describes the change in the apparent direction to a distant object caused by a change in the observer's location. In astronomy, it refers specifically to the very small change in the position of a star when observed from opposite sides of the Earth's orbit. This change, when measured, can be used to infer the distance to the star. The parallax of the nearest star, Proxima Centauri, is 0.768 seconds of arc.

Parsec
A unit of distance, often used by professional astronomers in preference to light years. A star at a distance of one parsec has a *parallax* of one second of arc. It is equal to 3.26 light years. The nearest star, Proxima Centauri, is 1.3 parsecs from the Sun. Distances within our Galaxy are generally expressed in kiloparsecs (1,000 parsecs; abbreviation kpc), whilst distances between galaxies are expressed in megaparsecs (1,000,000 parsecs; abbreviation Mpc).

Penumbra
This is the area of partial shadow around the main cone of shadow cast by the Moon during a solar *eclipse* or the Earth during a lunar *eclipse*. The term penumbra is also applied to the lighter and less cool region of a sunspot.

Perigee
This is the point in its *orbit* around the Earth at which an object is at its closest to the Earth.

Perihelion
This is the point in its *orbit* around the Sun at which an object, such as a planet, comet or asteroid, is at its closest to the Sun.

Planetary Nebula
Planetary nebulae consist of material ejected by a star during the latter stages of its evolution. The material thrown off forms a shell of gas surrounding the star whose newly-exposed surface is typically very hot. Planetary nebulae have nothing whatsoever to do with planets. They derive their name from the fact that, when seen through a telescope, some planetary nebulae take on the appearance of luminous discs, resembling a gaseous planet such as Uranus or Neptune. Probably the best known example is the famous Ring Nebula (M57) in Lyra.

Precession
The Earth's axis of rotation is an imaginary line which passes through the North and South Poles of the planet. Extended into space, this line defines the North and South Celestial Poles in the sky. The North *Celestial Pole* currently lies close to Polaris in Ursa Minor (the Little Bear), so the daily rotation of our planet on its axis makes the rest of the stars in the sky appear to travel around Polaris, their paths through the sky being centred on the Pole Star.

However, the position of the north celestial pole is slowly changing, this because of a gradual change in the Earth's axis of rotation. This motion is known as 'precession' and is identical to the behaviour of a spinning top whose axis slowly moves in a cone. Precession is caused by the combined gravitational influences of the Sun and Moon on our planet. Each resulting cycle of the Earth's axis takes around 25,800 years to complete, the net effect of precession being that, over this period, the north (and south) celestial poles trace out large circles around the northern (and southern) sky. This results in slow changes in the apparent locations of the celestial poles. Polaris will be closest to the North Celestial Pole in the year 2102, but it will then begin to move slowly away and eventually relinquish its position as the Pole Star. Vega will take on the role some 11,500 years from now.

Prime Meridian
The celestial prime *meridian* is the meridian on the sky that passes through the *vernal equinox*. It marks the zero point for measuring *right ascension*.

On the surface of the Earth, the prime meridian is the line of constant *longitude* which passes through the centre of the Airy transit telescope at the Royal Observatory at Greenwich in London. It was adopted by international agreement in 1884 as the origin for measuring longitude. Unlike the celestial prime meridian, it has no physical significance.

Prograde Motion
See Direct Motion.

Pulsar
This is a rapidly-spinning neutron star which gives off regular bursts of radiation.

Quadrature
This refers to the geometric configuration of the Sun, Earth and a *superior planet* when the elongation of the planet from the Sun, as seen from the Earth, is 90°.

Quasar
These are small, extremely remote and highly luminous objects which at the cores of active galaxies. They are comprised of a super-massive black hole surrounded by an accretion disk of gas which is falling into the black hole.

Reflection Nebula
See Nebula.

Retrograde Motion
A planet is in retrograde motion when its *right ascension* or ecliptic *longitude* is decreasing with the passing of time. This means that it is moving westwards with respect to the background stars. All *superior planets* undergo a period of retrograde motion around the time of *opposition*.

Right Ascension
The angular distance, measured eastwards, of a celestial object from the *vernal equinox*. Right ascension is expressed in hours, minutes and seconds.

Satellite
A satellite is a small object orbiting a larger one.

Seeing
The effects of atmospheric conditions on image quality experienced when carrying out visual observation and astronomical imaging of the night sky.

Shooting Star
The popular name for a *meteor*.

Sidereal Period
The time taken for an object to complete one *orbit* around another, measured with respect to a fixed direction in space.

Solar
Of or appertaining to the Sun.

Solar Eclipse
See Eclipse.

Solar System
The Solar System is the collective description given to the system dominated by the Sun and which embraces all objects that come within its gravitational influence. These include the planets and their satellites and ring systems, minor planets, comets, meteoroids and other interplanetary debris, all of which travel in orbits around our parent star.

Solstice
These are the points on the *ecliptic* at which the Sun is at its maximum angular distance (*declination*) from the *celestial equator*. The term is also used to denote the dates when the Sun passes these points on the ecliptic.

Spectroscope
An instrument used to split the light from a star into its different wavelengths or colours.

Spectroscopic Binary
This is a *binary star* whose components are so close to each other that they cannot be resolved visually and can only be studied through *spectroscopy*.

Spectroscopy
This is the study of the spectra of astronomical objects.

Star
A star is a self-luminous object shining through the release of energy produced by nuclear reactions at its core.

Star Clusters
Although most of the stars that we see in the night sky are scattered randomly throughout the spiral arms of the Galaxy, many are found to be concentrated in relatively compact groups, referred to by astronomers as star clusters. There are two main types of star cluster – open and globular. Open clusters, also known as galactic clusters, are found within the main disc of the Galaxy and have no particularly well-defined shape. Usually made up of young hot stars, over a thousand open clusters are known, their diameters generally being no more than a few tens of light years. They are believed to have formed from vast interstellar gas and dust clouds within our Galaxy and indeed occupy the same regions of the Galaxy as the nebulae. A number of open clusters are visible to the naked eye including Praesepe (M44) in Cancer, the Hyades (C41) in Taurus and perhaps the most famous open cluster of all the Pleiades (M45), also in Taurus.

Globular clusters, as their name suggests, are huge spherical collections of stars. Located in the area of space surrounding the Galaxy, they can have diameters of anything up to several hundred light years and typically contain many thousands of old stars with little or none of the nebulosity seen in open clusters. When seen through a small telescope or binoculars, they take on the appearance of faint, misty balls of greyish light superimposed against the background sky. Although some form of optical aid is usually needed to see globular clusters, there are three famous examples which can be spotted with the naked eye. These are 47 Tucanae in Tucana, Omega Centauri in Centaurus and the Great Hercules Cluster (M13) in Hercules.

Star Colours

When we look up into the night sky the stars appear much the same. Some stars appear brighter than others but, with a few exceptions, they all look white. However, if the stars are looked at more closely, even through a pair of binoculars or a small telescope, some appear to be different colours. A prominent example is the bright orange-red Arcturus in the **constellation** of Boötes, which contrasts sharply with the nearby brilliant white Spica in Virgo. Our own Sun is yellow, as is Capella in Auriga. Procyon, the brightest star in Canis Minor, also has a yellowish tint. To the west of Canis Minor is the constellation of Orion the Hunter, which boasts two of the most conspicuous stars in the whole sky; the bright red Betelgeuse and Rigel, the brilliant blue-white star that marks the Hunter's foot.

The colour of a star is a good guide to its temperature, the hottest stars being blue and blue-white with surface temperatures of 20,000 K or more. Classed as a yellow dwarf, the Sun is a fairly average star with a temperature of around 6,000 K. Red stars are much cooler still, with surface temperatures of only a few thousand K. Betelgeuse in Orion and Antares in Scorpius are both red giant stars that fall into this category.

Stationary Point

A planet is at a stationary point when its motion with respect to the background stars changes from **direct** (motion) to **retrograde** (motion) or vice versa. All **superior planets** pass through two stationary points at each **apparition**, once before **opposition** and again after opposition.

Superior Planet

A superior planet is a planet that travels around the Sun outside the **orbit** of the Earth.

Supernova

Supernovae are huge stellar explosions involving the destruction of massive stars and resulting in sudden and tremendous brightening of the stars involved.

Synodic Period

The synodic period of a planet is the interval between successive **oppositions** or **conjunctions** of that planet.

Transit

1 – The instant when an object crosses the local **meridian**. When the object's **local hour angle** is zero, this is known as upper transit, and marks the maximum **altitude** of the object above the observer's **horizon**. When the object's local hour angle is 12 hours, it is known as lower transit.

2 – The passage of Mercury or Venus across the disk of the Sun, as seen from the Earth, or of a planetary satellite across the disk of the parent planet.

Umbra
This is the main cone of shadow cast by the Moon during a solar *eclipse* or the Earth during a lunar *eclipse*. The term umbra is also applied to the darkest, coolest region of a sunspot.

Variable Stars
A variable star is a star whose brightness varies over a period of time. There are many different types of variable star, although the variations in brightness are basically due either to changes taking place within the star itself or the periodic obscuration, or eclipsing, of one member of a *binary star* by its companion.

Vernal Equinox
The vernal equinox is the point at which the apparent path of the Sun, moving from south to north, crosses the *celestial equator*. In the Earth's northern hemisphere this marks the start of spring, whilst in the southern hemisphere it is the start of autumn.

Zenith
This is the point on the *celestial sphere* directly above the observer.

Zodiac
The band of *constellations* along which the Sun appears to travel over the course of a year. The Zodiac straddles the *ecliptic* and comprises the 12 constellations Aries, Taurus, Gemini, Cancer, Leo, Virgo, Libra, Scorpius, Sagittarius, Capricornus, Aquarius and Pisces. The ecliptic also passes through part of the constellation of Ophiuchus, as delimited by the boundaries defined by the *International Astronomical Union*, but Ophiuchus is not traditionally considered a constellation of the Zodiac.